Species and Machines

This book offers a re-examination of the relationship between humans and nature with a new methodology: by examining our entanglement with machines. Using central ideas of critical theory, it uncovers the suppression of nature through technology, tools and engines. It focuses on the ways in which human social forms have actively subjugated and destroyed other species in order to enhance their own social power and accumulation, leading to a new Anthropocene epoch in which human intervention is signalled in the geological record.

Beginning with an account of the interactions between humans and other species, the book moves on to explore the hidden history of Marx and his obsession with machines, as well as new attempts to rethink a Marxist ecology, before proceeding to examine the manner in which technologies were used to suppress and destroy one particular species – the whale of what we call the Cetacean holocaust. Following this, there are analyses of the emergence of the 'human encampments' of the cities and the rise of mobile, locomotive cultures, and consideration of the relationship between machines of memory, and the 'capturing' of nature.

A radical rethinking of classical social theory that develops new ways of thinking about ecological catastrophe and nature, this book will appeal to scholars of social theory and environmental sociology.

Martyn Hudson is an associate researcher in the School of Arts and Cultures at Newcastle University, and author of *The Slave Ship, Memory and the Origins of Modernity*, and *Ghosts, Landscapes and Social Memory*.

Species and Machines

The Human Subjugation of Nature

Martyn Hudson

Routledge
Taylor & Francis Group
LONDON AND NEW YORK

First published 2018 by Routledge

2 Park Square, Milton Park, Abingdon, Oxfordshire OX14 4RN
52 Vanderbilt Avenue, New York, NY 10017

Routledge is an imprint of the Taylor & Francis Group, an informa business

First issued in paperback 2018

British Library Cataloguing-in-Publication Data
A catalogue record for this book is available from the British Library

Library of Congress Cataloging-in-Publication Data
Names: Hudson, Martyn, author.
Title: Species and machines : the human subjugation of nature / Martyn Hudson.
Description: Abingdon, Oxon ; New York, NY : Routledge, 2018. | Includes bibliographical references and index.
Identifiers: LCCN 2017014270 | ISBN 9781138087972 (hbk) | ISBN 9781315110165 (ebk)
Subjects: LCSH: Technology–Social aspects. | Technology and civilization.
Classification: LCC T14.5 .H8335 2018 | DDC 304.2/8–dc23
LC record available at https://lccn.loc.gov/2017014270

ISBN: 978-1-138-08797-2 (hbk)
ISBN: 978-0-367-20820-2 (pbk)

Typeset in Times New Roman
by Integra Software Service Pvt. Ltd.

Contents

Acknowledgements

In some ways this book is an extended commentary on themes first outlined by Theodor Adorno and Max Horkheimer in *Dialectic of Enlightenment* as a response to the civilisational crisis initiated by European fascism and dictatorship. Their thesis was fundamentally concerned with the human domination of nature accelerated by the Enlightenment project. In recent years the project of critical theory has been sustained by the Heathwood Institute, a project which has now reached its end. I would like to thank R.C. Smith and Heathwood for acting as the laboratory for the ideas in this book. The critique of the Enlightenment project from within itself is still of urgent importance in the face of new waves of endarkenment and of what Victor Serge once called 'midnight in the century' (see Weissman 2001).

First, I would like to thank Neil Jordan and Shannon Kneis for their huge support with this and previous projects. I would like to thank my colleagues and friends who have contributed to ideas about history and prehistory, literally on the ground: Emily Hesse, Ben Reche, Michal Yankowski, Gavin Parry, Michele Allen, Mike Woolley, Dan Seale, Nick Brereton, Mick Garratt, Keith Wilson, Bob Howe, Alistair Bowden, Savita Sathe, Ian Hodgson, Dave Allison, Bill Pennell and the entirety of our 'Helly' group. I owe a debt of thanks to Culture Lab and other colleagues at Newcastle University and specifically to John Bowers, Ben Jones, Jane Dudman, Alexia Mellor, Eric Cross, James Davoll, Frances Rowe, Ben Freeth, Gabi Arrigoni, Tom Schofield, Rob Airey and Eric Cross. Many of the ideas in this book have their origin in conversations with Julie Crawshaw and her work on art and design history and ethnography. During the period of writing I have been involved with a project called 'Landscape Archaeology of the Great Depression in the North-East of England' based at Durham University and I would like to thank David Petts, Gwilym Williams and Quentin Lewis for their ideas, conversations and materials. Quentin's work on the landscape archaeology of the United States led to important conversations at a decisive moment.

My other project during this period was that based at the Newbridge Project, Newcastle. I would like to thank David Butler, Jessie Joe Jacobs, Charli Gregory, Rebecca Huggan, Paul Richter, Ed Wainwright, Julia Heslop, Sarah Tulloch, Taryn Edmonds and the Newbridge members for their ideas and including the Newcastle Institute for Creative Arts Practice. The temporary 'human encampment' of Newbridge provided respite from the solitude of researching in northern latitudes. I would also like to thank Sacha Ismail, Richard Gold, David Hirsh, Eric Lee, Martin Thomas, Ed Whitby, Richard Greeman, John Berger, Jean Mohr, Cathy Nugent, Marcus Rediker, Gwendolyn Midlo Hall, Suzi Weissman, James Bloodworth, Pete Cookson, Alan Johnson and others in the social struggle over history and its future course. I would also like to thank Peter Linebaugh and Jason Hribal for their dual understanding of thanatocratic regimes wielded over humans and animals. It strikes me that Jason's work on human–animal proletarian solidarities marks an important moment in critical animal studies. The section on the theatrical performance of 'flensing' in Chapter 3 was suggested by being part of a performance by Chris Goode and his collaborators at *Redrawing the Maps*, a conference on John Berger in 2012. Thank you to Chris and to Dougald Hine. I would like to thank the staff and volunteers at Chetham's Library, Manchester who helped me explore the alcove, looking on to the slums that so galvanised Engels, where Marx and Engels would sit and talk and develop the foundational ideas of historical materialism. I would also like to thank Whitby Literary and Philosophical Society and Whitby Museum, Pannett Park for shaping my understanding of the Scoresbys and of the whaling industry.

On Holocaust studies and memorialisation, I would like to note the work of my dear friends Yael Reicher and Emmanuel Tzwern and some of the stories they have shared with me about their families and the Les Mazures camp in the Ardennes. One element of the work presented here is due to a dialogue with Theodora Polenta.

This book is dedicated to Olivia Hudson-Carder, my dearest friend and comrade and my daughter.

Preface

The multiple revolutions of the eighteenth century and the acceleration of the human domination of nature created an intellectual crisis about the nature of human beings themselves. If humans were essentially products of nature then the self-destruction or abandonment of their natural habitats was catastrophic. This catastrophe was both external in their relations to other human beings and also internal in their relations to themselves. The emergence of sociology was an attempt to both understand and rectify this state of 'exile' and transformation. The intellectual tumult of the Enlightenment across Europe, the revolutionary prefiguration of the commons and mass democracy in France and the industrial acceleration in Britain initiated those moments of reflection that we find in the genesis of classical sociology. Of course, the intellectual torment and reflection of human beings about their very status as human beings as divorced from nature was not a new phenomenon. Indeed it is present in classical Hellenic civilisation, in medieval Islam and further back into Palaeolithic cave art. However, the acceleration of certain tendencies and social dislocations were indeed creating a sense of crisis: the forced captivity and migration of millions of Africans across the middle passage to the plantation regimes of the new world, the 'flight of the serfs' into the factories of the industrial city, the geographical and scientific expansion of exploration, the new categories and lineages of human beings initiated by science and, ultimately, by Darwinian evolution, the fall of kings and the emergence of mass democracies, warfare and capital accumulation, the potential supersession of religious forms. Modernity was indeed transformative and traumatic for those experiencing and enduring it. Adorno and Horkheimer's *Dialectic of Enlightenment* pursued a radicalised enlightenment project just as much as it condemned aspects of social and natural domination that it heralded. C.L.R. James, writing his magnificent book on Melville and whaling at the same time as Adorno and Horkheimer were writing their *magnum opus*, saw in Ahab the new Promethean 'totalitarian type', subjugating species in a relentless search for knowledge and destruction.

For classical sociology the acceleration of these tendencies towards exploration, expansion and accumulation created a profound internal crisis for human beings; about what they were, where they were meant to reside, what was happening to them in the factories, why they were being transported, how they were being transformed into commodities. Nowhere is this more marked than in the human relationship to 'nature' and to other species. The emergence of new technical civilisations were leading to new relationships between social forces and what they could extract from the earth or capture from the sea. Indeed, whole civilisations and epochs were built upon extraction and accumulation. The internal dislocations that those accelerations imposed upon human beings were of great magnitude. At the same time the anthropocentric and accumulative human 'dominion' over nature and its resources created new ecologies and had disastrous consequences for habitats and other species. Habitats were annihilated, species forced into extinction. Geographical transformations meant the transport of human populations in terms of mass migration. The development of the extraterritoriality of species which had their origin in specific spaces became part of both the world-economy and of zoological display.

The demarcation, or 'vivisection', between humans and other animals separated humans out from the animals as their masters, separated some parts of humanity away from the labouring mass, and relegated large parts of humanity itself to subhuman or non-human status. New polarities were constructed between animals and humans (often sanctified by either God or naturalist) and new continuums of species generated in which they were considered here more human, there less human. In a sense, the very value of a species rested either upon its similarity to humanity in terms of shared characteristics, or upon the capacity of the species to be subordinated and subjected to human will, or for a species to be the 'raw' materials of an industrial civilisation and used in exactly the same way as, for example, mineral deposits. The machines that humans generated prosthetically to subject nature to their will also ended up subsuming large parts of humanity itself. New plebeian classes were created that contained both humans and animals. The sociological project to elucidate the relation between humans, other species, nature and machines became of urgent necessity for its initial practitioners.

This book is an attempt to account for that disaster, not in terms of any attempt at restoral or restoration of nature or some golden age of primitive humanity for that is impossible. It is an attempt to rethink our basic sociological categories of humans, species and the machines that we have created to subsume nature. This means returning again to the origins of social theory and its attempt to think about *catastrophe.* A certain number of catastrophes led intellectuals to rethink their basic categories of nature and the human; Voltaire and the Lisbon Earthquake, Rousseau and the destruction of the 'savages', Engels and the degradation of Manchester – that first city of industrial civilisation.

Much was unwritten of course that we might hope had been written. Circumstances of history and geography often precluded ways of thinking about animals that might seem of necessity to us now. But the warnings of disaster were there. What accounts then for our return to Marx and others, to political economy, to an emerging anthropology? First, we need to write a history of modernity from the standpoint, as Walter Benjamin has said, of *catastrophe*. But this is a multiple disaster; for humans, for species, for ecologies. Second, the early sociologists believed that their description of phenomena also entailed a social and political programme. If there is a social catastrophe then there might be programmes, resolutions, knowledges that might ameliorate or suspend the conditions of the social forces that initiated the crisis. Third, we often talk of humans, other animals and machines as isolated categories. In reality it is their entanglement and entwining that is both catastrophic and holds out the promise of whatever restoration of natural form is possible. If it is at all possible.

The persisting idea that it is modernity that problematises human identity presupposes the idea of the human as a 'natural' being forced into conditions of strangeness and exile. This of course is only part of the truth. The sustained reflection on identity, dislocation and interior being has its origins in deep antiquity. Modernity and capitalism have simply accelerated the dislocation. Further, the human 'dominion' over nature develops the conditions not just for the extinction of other species but potentially of humanity itself. Whether one sees acceleration and human liquidity as the consequence of the Neolithic agricultural revolution and that intervention in nature or in the development of the steam engine, it is the case that humans have indelibly marked the organic and inorganic conditions of the earth. The signals of human intervention in the geological record have already initiated some geologists to rethink our epoch as the Anthropocene. Related to this is the extension of the 'neobiota' of the human, indeed particular forms of human, into every aspect of nature. Favoured species extend their geographical reach while others are suppressed. Often this is because they offer modes of extraction either in their flesh or their labour power, or in other cases because they are either a perceived threat to humanity or their habitats have been destroyed by relentless modernisation.

In the first chapter we begin to elucidate a social theory of 'machination' and accumulation and the entanglement of humans and their machines with nature. The destruction of species and their diversion away from 'natural' forms of progression by human intervention is situated within new concepts of 'nature'. This also concerns the relation of human beings to 'nature' and their relation to themselves as 'natural' beings. We also introduce in this chapter the idea of transhumant humanity – the humanity of migration, movement and pastoralism as a way of thinking about the potentialities of human and animal development.

In the second chapter we look specifically at Marx and his documentation of the accumulative and productive machines of capital. In some ways the most remarkable passages of Marx lie in the marginalia of his writings and in that of Engels. We examine specifically Marx's notebooks from the London winter of 1857–1858 collectively known as the *Grundrisse*. Within this work there is a 'fragment' on machines but also the development of a concept of 'metabolism' which has become central to new thinking about ecology – specifically in the social theory of John Bellamy Foster and his collaborators. This chapter extends that thinking about 'metabolic rifts' between humanity and nature in new directions and aspires to rethink the Marxist programme for and over nature.

In the third chapter we examine again the entanglement between species, nature and machines by uncovering the hidden history of a specific machinic industry – the historical 'animal genocide' of the 'Cetacean holocaust'. The rift between human and nature is nowhere more apparent than in the mass destruction of the whale populations of the North Atlantic in the seventeenth, eighteenth and nineteenth centuries by the weaponised machines of whaling ships. Not only did this take place on a mass-industrial scale, the 'fruits' of the massacres were central to the development of the factory system. At the same time as the science of the Enlightenment was uncovering and classifying a whole taxonomic system of Cetaceans (including dolphins), the wholesale slaughter of the whale populations created 'illuminants' for homes and factories. Its oil greased the machinery of the great manufactories, its labouring sailors became part of the first Atlantic proletariat. The entanglement of Cetacean metabolism with the human species and their machines was both catastrophic and a symbol of a human dominion over nature which is now collapsing.

In the fourth chapter we examine the extension of the 'camps' of humanity; its encampments, domains and cities. Examining first the 'imaginaries' of the urban we look at the strata and sedimentation of the human 'occupation' of nature and territory. The human 'camp' was often an enclave which was mobile and in movement across territory either for pastoral or military purposes. The development of permanent occupations transformed human cultures and their relationship to nature and other species. The colonisation of territories and their organic inhabitants has often led to 'speciescide' as animals are slaughtered to service the consumption of metropolitan human populations. The industrial slaughterhouses of the cities have led some species to the brink of, or actual, extinction. The permanent human occupation of urban spaces has led to new social relations between humans, species and nature and new forms of entanglement.

The extension of human power over geology and geography is discussed in Chapter 5 particularly in relation to the 'locomotive' machinic cultures of the earth. The idea of bounded territories has been challenged by a global, mobile culture – specifically and initially that of the steam locomotive. Intimately related to empire and colony, the train was a machine for the transmission of populations and labour, of commodities and raw materials and of armies. This chapter looks at the cultural history of the development of locomotive power and its relationship to animals and the natural world that the 'engine' traverses.

In Chapter 6 we return to what Bernard Stiegler has called the 'third' memory of machines. Not only is memory encased in the artefacts of human labour and scored upon the natural strata of the earth, machines can also be part of archiving processes. Archiving nature, species and natural forms is itself imbued with 'archival power', classification and measurement. Exploring the relationship between natural archives, human observation and machines might help us to elaborate ways of evading the human-centric observations of nature.

In Chapter 7 we conclude by summarising the discussion and looking at the potential destinations of what I call the 'human movement'; its limitations, migrations, extinctions, crimes and the idea of the restoration of nature and the 'wild'. The potential displacement of the human animal from the pinnacle of the biological and evolutionary hierarchy and indeed the dissolution of that hierarchy has opened up new possibilities for human–animal–nature relations. It also signals a warning that the interventions into nature that began with the Neolithic revolutions (and before that with the very birth of the *idea* of the human in the Palaeolithic caves) and then with the obsession of the classical world with machinery may have to dramatically alter. The maintenance of those systems of metabolic and organic exploitation and the extraction and accumulation of the resources of inorganic nature cannot be sustained.

The novel rethinking of Critical Theory, within which this work is situated, must be performed in conjunction and collaboration not just with other theorists, but with other natures and other species. The occasion for the development of these ideas was in the aftermath of Philip Low and his collaborators' *Cambridge Declaration on Consciousness* at the University of Cambridge in July 2012 which marks a transformation in the scientific study and exploitation of animals (Low 2012). Its insistence that neither humans nor animals are 'biological machines' and its rejection of taxonomies and classifications of human and animal separation and difference have brought the actuality of the 'sentient' earthly being back to the forefront of scientific practice and theory. The entanglement of biological species and machines is more significant however, than the emergence and rejection of the idea of the 'biological machine'. The projects either to separate organic beings from mechanical analogies or to see the essential identity of being and machine, both miss the point that from the very beginnings of recorded history the organic and the machinic have been seen as both allies and contestants. Whatever the worth of those analogies the cultural, economic and scientific effects of human–animal–machine relations have been overwhelming.

Theorising the relationship between different species, nature and machines has to be in its very nature historically and geographically oriented. The extension of anthropocentric domination across territory and emerging and coalescing in different historical periods entails an understanding of the material production and forces of the new human dominions and 'imperialism'. But theory also allows what Johannes Fabian has called 'a world of interest and amazement, of desire and pleasure, of involvement and performance, of people who move in celebratory processions' (2001: 6). Theory can experiment, prefigure, play and bring into concatenation different concepts, historical forces and discourses – above all it is in Derrida's sense – *critical* (1995: 357). This critique itself can be understood as historically formed discourse and therefore deconstructed but it 'must never be renounced' (1995: 357). The very deconstructibility of discourses is not a disqualification or a disavowal but of doing critiques of critiques – 'of thinking its possibility from another border' (Derrida 1995: 357).

In the light of the successive human transgression of borders and boundaries and their destruction of habitats and species this thinking from across 'another border' about our own critiques and concepts, as both human beings and more specifically theorists who commentate on the human 'social', is absolutely integral any account of the history of technology and domination. Ultimately as Blanchot has said, in his writing of the disaster – 'It is dark disaster that brings the light' (1986: 7). The luminescence of the Enlightenment project of understanding should not be abandoned but critiqued and extended. While persisting in the act of illumination one must recognise that the illuminants of both the factory system and the lights of scientists, explorers and taxonomists were themselves based upon the almost wholesale destruction of a species. That history and that extension over geography cannot be undone but understanding what they have cost the earth and humanity can help us elaborate new routes into a future of cohabitation rather than of dominion.

The first machines were in fact an entanglement between the organic and the inorganic – wooden railways, the 'wooden world' of the great ships, tools made of bone and so on. Increasingly the human inventors of the machine recognised that not only did the machine have to interact with nature and be bound by natural laws, geologies and geographies but that organisms offered templates for the form of machines themselves. The replicability of organic forms in flight machines, in the 'veins' of movement and transportation and in automata modelled upon humans, witnesses both to entanglement between the organic and inorganic. It offers also potential ways of mending the rift and the drift between different forms of metabolism. This might happen by way of new ways of thinking about the 'monstrous' products of hybrid beings by nature and by human invention. This might mean the extension of the technological prosthesis of the human, or a combination between the human and other species, indeed it might entail a supersession of the human and new 'monstrous' evolutionary routes. Transhumanism offers potential dangers to other species in those new concatenations and conglomerations of species-combination. At the same time the 'machine' might offer potential solutions to the metabolic rift between species and between humans and nature.

1 Introduction: Nature, species and machines

In this chapter, we begin to elucidate a social theory of 'machination' and accumulation and the entanglement of humans and their machines with nature. The destruction of species and their diversion away from 'natural' forms of progression by human intervention is situated within new concepts of 'nature' and the relation of human beings to it and their relation to themselves as 'natural' beings. We also introduce in this chapter the idea of transhumant humanity – the humanity of migration, movement and pastoralism as a way of thinking about the potentialities of human and animal development. In some ways it is impossible to separate human beings from 'technics'. The emergence of *Homo sapiens* was intimately related both to aesthetic representation and the transformation of natural artefacts, like stones, into tools. The machine is not simply a phenomena of modernity and industrialisation but a technical prosthesis of human beings which has accompanied each epoch of their development albeit in different forms.

The mechanic and the organic are often syncretic. This is seen at its most stark with the development of digital cultures where new hybrid forms of human and machine are in the process of development. But the mechanical has a history going back into classical antiquity and indeed syncretic automata have their origins in ancient Greece. This notion of classical hybrid beings has been explored elsewhere (Hudson 2016a), as have the ways in which whole captive populations have had a particular type of syncretism industrially imposed upon them (Hudson 2016b). Suffice to say that inscribed into human cultures are obsessions with dominating nature, extracting from it, and transforming these forms of accumulation into more refined and extensive machines to dominate nature even further. The processes of machine acceleration and accumulation might find their apotheosis in the factories of Salford and Manchester that Friedrich Engels surveyed in the 1840s but their genesis and subsequent modes of generation have deeper histories.

The accumulation of the 'stones' of inorganic nature to make tools have their counterpart in the extraction from organic nature – both its 'fruits' and its 'flesh'. The vast industrial slaughterhouses of Chicago were certainly at one end of a continuum but the agricultural revolutions of the Neolithic and the tranhumant pastoral and hunting cultures of the early Mediterranean world were also accumulative if at the other end of that continuum of natural intervention (Braudel 1975: 85–95). What Braudel calls the 'continual migrations' of transhumance and nomadism were part of the vast micro-economic systems of the Mediterranean world (1975: 85). The two central types of transhumance were 'normal' migrations from the plains to the highlands and 'inverse' migration – that from the highlands to the plains. These seasonal winter and summer transhumances could be observed throughout the littoral of the Mediterranean. As Braudel says: 'Transhumance implies all sorts of conditions, physical, human, and historical' (1975: 87). Studying these pastoral migrations for Braudel might imply slow, evolutionary development of these forms of movement but they can also be the product of 'catastrophe' and 'revolution' (1975: 88). What Braudel's model does show is that humans are often nomadic and migratory and that the idea that they are fixed in pre-capitalist economic formations to specific territories and even fields is misguided. Perhaps more decisively the transhumants were not alone – their very purpose in seasonal movement were their 'flocks' and 'cattle'. This was the beginning of what would become the transhumant passage of whole species of animals across plains and mountains and into the industrial slaughterhouses of the metropolitan occident.

The domination of territories and the passages through them was also about the domination of non-human species and indeed of those humans who were not considered part of that species. This involved the 'mastering' of certain historical narratives about history and geography in terms of lineage and ownership, rights of traverse, and taxonomic classifications about who and what could be subjected to human, or specific human, rule and mastery. If certain types of human beings were to eventually become 'appendages' to the machines of the industrial metropolis, so were non-human species seen as appendages, or even sometimes as fuel. This fuel was mechanically accumulated from the carcases of non-human species in order to fuel the machines themselves. Often this was quite literal, as in whale oil to fuel machines, but also metaphorically in terms of the use of 'flesh' to fuel human consumption and thereby further extension and penetration into 'nature' and other beings than themselves. This process found its apotheosis in the industrial human death camps of Treblinka and elsewhere but some commentators see the human suppression of animals as itself an 'eternal Treblinka' (Patterson 2002).

Technics has always modelled itself upon the templates of nature. Often machines themselves have been products of organic nature built from its material. The machine itself without the inorganic would be an impossibility. It is increasingly the case that in order to somehow accrue the capacities that organic entities have as part of their being, humans have shaped machines in their image. The use of their lines, patterns, natural geometries and the replication of their natural 'mechanics' are leading to new machinic membranes being built around organic beings – exoskeletons which rather than replicate and copy are actually shaped around the natural phenomenon. The human aspiration to fly and its multiple failures to do so have continually obsessed engineers and inventors. The Soviet sculptor and scientist Vladimir Tatlin used models of animals in order to design and produce new types of flight machines and there are new prosthetic examinations of how automata can be modelled on insects and bacteria in order to produce 'electric sheep'.

Understanding the entanglements between different species, nature and machines means understanding their moments of contact. It also entails understanding their gestures of reconciliation. Understanding their contests and their aggressive manoeuvres means understanding the social world as itself impossible without either animals or machines. It also means that while we reaffirm the essential contiguousness between humans and animals we must also negate the anthropic renditions of the 'humanised' animal. It means toppling humanity both from its taxonomic pinnacle at the tip of creation and disarming the human military onslaught against other species. Not that this exercise is about telling us more about *us* by way of animals and understanding us as natural beings. Rather, it lies in understanding that entanglement could contribute to new social coalitions in which nature, human and species are not opposing forces, but allies in a coalition to reform if not restore our ecologies. Opening ourselves up to the idea that humans are porous and that our relations with others (including species and machines) define us might mean redefining our humanity. As Roberto Esposito has said in his reading of Heidegger:

> ...the animal is not the ancestral past, the stone guest, the mute enigma, but the future of man: it is a place, and a threshold, from which man can only be stimulated in view of a more complex and open elaboration of his *humanitas.*
>
> (Esposito 2011: 84)

The significance of The Cambridge Declaration on Consciousness

In itself, a declaration by a group of Cambridge scientists on 'neurological substrates' might seem to be relatively underwhelming. In actuality the emergence of the declaration, initiated and written by Philip Low, sustains a profoundly important set of ideas on human–animal relations and what has come to be known as 'critical animal studies'. On 7 July 2012, at the Francis Crick Memorial Conference on Consciousness in Human and Non-Human Animals, at Churchill College, Cambridge, a number of scientists met to discuss the neurological substrates of consciousness and signed what would be known as *The Cambridge Declaration on Consciousness.* They noted that there was an inability of non-human animals, and indeed some humans, to communicate about and from their 'internal states'. Low and his fellow practitioners noted the emergence of new kinds of data which are emerging from human and animal studies including new (in humans) 'non-invasive techniques' to survey aspects of consciousness. As Philip Low and his peers note:

> The neural substrates of emotions do not appear to be confined to cortical structures. In fact, subcortical neural networks aroused during affective states in humans are also critically important for generating emotional behaviors in animals. Artificial arousal of the same brain regions generates corresponding behavior and feeling states in both humans and non-human animals. Wherever in the brain one evokes instinctual emotional behaviors in non-human animals, many of the ensuing behaviors are consistent with experienced feeling states, including those internal states that are rewarding and punishing. Deep brain stimulation of these systems in humans can also generate similar affective states.
>
> (Low 2012)

For Low, 'near human-like' levels of consciousness can be detected in birds evincing homologous neurological states to humans and 'Magpies in particular have been shown to exhibit striking similarities to humans, great apes, dolphins, and elephants in studies of mirror self-recognition', (Low 2012). The evolutionary development of consciousness in animals is parallel to humans and certainly comparable and analogical. The Declaration then states in conclusion:

> The absence of a neocortex does not appear to preclude an organism from experiencing affective states. Convergent evidence indicates that non-human animals have the neuroanatomical, neurochemical, and neurophysiological substrates of conscious states along with the capacity to exhibit intentional behaviors. Consequently, the weight of evidence indicates that humans are not unique in possessing the neurological substrates that generate consciousness. Nonhuman animals, including all mammals and birds, and many other creatures, including octopuses, also possess these neurological substrates.
>
> (Low 2012)

The significance of the Declaration lies in five factors that it intimates or makes possible. First, an understanding resting on neurological data that there is a convergence or a parallel between human and animal consciousness and affective states. Second, the displacement from its pinnacle of human consciousness as something distinctive from the entirety of the rest of animal life. Third, the intentionality of animals in terms of consciousness and intimating also the convergence with human intentionality and all that that may imply. Fourth, the implication of understanding the neurological substrate using 'non-invasive' techniques and extending this to animals (who converge with humans in their affective states and intentionality). Fifth, the notion of 'sentience' and the huge implications for understanding large realms (or kingdoms or republics) of animal consciousness, specifically whales as we shall see, that cannot be accessed or properly understood by human 'invasion'. The displacement of human beings from their hierarchical eminence and the reassertion of animal sentience is part of a continuing process to question the human intervention into and invasion of nature. In the reassignation of geological epochs by looking at the nature of human intervention as signals in the geological record we can display the epochal and continental machination of human beings in those records. This has led some scientists to develop the concept of the Anthropocene geological epoch as a way of thinking about the catastrophic interventions of human beings into biology and the structures of the earth.

Anthropocene cultures and signals

Mark Williams and his collaborators have argued that,

> *Homo sapiens* is itself a neobiotic species par excellence, and its members have left their remains and also produced a set of characteristic (and potentially fossilisable) anthropogenic structures that extend from ancient Mesolithic campsites, through the ancient cities of Mesopotamia, to the modern research bases of Antarctica.
>
> (Williams et al. 2015: 203)

The signals of human intervention that appear in the geological (fossilisable) record demonstrate the power of a specific type of 'Neobiota' – that of the human and its capacity to extend itself and its favoured species globally. It also records the damage to non-human habitats and to unfavoured species. The development of the 'Anthropocene' geological concept was first suggested by Crutzen and Stoermer (2000) but it is now achieving a huge discursive impact in geological circles (Barnosky 2014, Haff 2014a, 2014b, and Schwägerl 2014). Recent work by Stoner and Melathopolous (2015) in critical theory has pointed to the power of the concept both in terms of illustrating the catastrophe of human intervention but also observing the pessimism about 'restoration' that might lie at the heart of the Anthropocene discourse. As has been noted elsewhere:

> The development of new ways of periodising geological time has been predicated on understanding the global signals of human intervention in the geological record. This intervention is largely the result of two related processes – the expansion of human organisation and sociality and the expansion of human agency into the development of technology – specifically machines. The global technosphere produced by the social world of human beings now parallels and threatens to subsume the global biosphere. In fact, the biosphere itself now *interacts* in complex ways with human sociality including the geographical extension of some species and the extinction of others. The interactions between the human and non-human world are often themselves part of complex productions of machines: machines of extraction, production, and knowledge. The development of mechanics has been enmeshed in ideas of using machines to enhance human power as prosthetics or extensions of the human body or the social formation. Using an idea

> from recent developments in technology we can think of these machines as social machines whether in computing or factories or slave ships. But these machines are no longer simply to be thought of as mechanical. Not only can we use biomorphic analogies in order to understand machines, new machines are being built which rest upon biological imperatives and foundations rather than simply mechanical ones.
>
> (Hudson 2015: 1).

The interaction between different species, machines and the natural world may indeed support the redefinition of geological epochs and particularly mark the acceleration and accumulation of the human species. If one central marker of the 'anthropocentric' is that signals of human intervention appear geologically then what kinds of signals are these? First, the impact of humans upon meteorological conditions, climate and pollution should be displayed clearly in the record. Second, the extinction of certain types of flora and fauna should be revealed. These may be because of a direct intervention by humans to burn and destroy certain types of habitats like forest to thereby increase the human accumulative capacity. Equally, it may be the extinction of fauna through hunting and persecution either for food and fuel or to protect the human enclaves from the threat of contestatory species. Third, interventions in water and oceans should be signalled, not just in terms of the extinction of certain types of flora and fauna perhaps as a consequence of human-initiated changes in meteorological conditions, but also in terms of the warming of the oceans, the loss of natural ice-habitats and the flooding of geological structures which hitherto, had been untouched. Fourth, as we shall see, the anthropocentric diminishment of species through taxonomy and classification creates a form of speciation in which the being of other species is submerged into hierarchical systems which facilitate human dominion and hunting for pleasure, food or fuel. These taxonomic discourses not only have social effects but also geological effects in terms of absence from fossil records or material remnants of the dead and even extinct species. Fifth, aside from the fuel which is needed to power the vast global metropolitan areas, the 'metropoli' of the urban and the city have extended themselves over huge territories of plain, desert and forest. Unwilling to merge and cohabit with natural forms the cities, often containing millions of human beings, have extended mechanics imperialistically over the organic. In some ways this, rather than a signal in the geological record, means the extinction of the geological record itself across those territories.

The material artefacts and transportation systems of the cities have been attractive loci for a mobile humanity which has largely severed itself from nature in all ways apart from its exploitation. Donna Haraway has noted the 'planetary' consequences of the Anthropocene and our relations and interactions, often dominatory, with other species:

> There is no question that anthropogenic processes have had planetary effects, in inter/intra-action with other processes and species, for as long

> as our species can be identified (a few tens of thousand years); and agriculture has been huge (a few thousand years). Of course, from the start the greatest planetary terraformers (and reformers) of all have been and still are bacteria and their kin, also in inter/intra-action of myriad kinds (including with people and their practices, technological and otherwise).
>
> (Haraway 2015: 159)

The analogies with the machine as bacteria, insect and swarm would itself become part of the mythology of the modern machine. The rise of the machine might find its ultimate expression in the metropolis but its origin is archaic.

The rise of the machines

> The figures of Daedalus, automata, like slaves, existing at the bidding of their master and maker, will always flee if they can.
>
> (Page duBois 2003: 248n)

Page duBois hints that these automata, enslaved to their masters as they are, are constantly trying to escape domination and evade their destiny as simply 'prosthetics' rather than independent beings. Machines, specifically those which are modelled on organisms, have their origin in deep antiquity. Classical civilisation was obsessed by both the mythological origins of machines and their makers and in making them; wooden horses, automata, musical instruments, clocks (Berryman 2009, Bromley 1986, Bruce 1913, Drachmann 1963, Koetsier 2001, Truitt 2015). Making was often entwined not just with machines but with aesthetics, with Daidalos, the originator of the labyrinth, automata and prosthetic flying wings, the original craftsperson and designer (Morris 1992). These artefacts were machines based not just on the observation of nature and organic beings but specifically designed to intervene in history and society either to conquer its armies, make it more musical or to protect their makers. Further, they were also extracted from nature and interventionist in nature itself – specifically designed to be syncretic and hybrid in their combination of the organic and non-organic. The wings of Icarus are the most powerful elaboration of this hybrid, prosthetic entity and many of these machines reflected the obsession that Ionia and classical civilisation had with hybrid beings like Centaurs and indeed gods (Hudson 2016a). As Brumbaugh has said in his now classic account of the ancient machine:

> The designers of automata seem to have become progressively more ambitious, and their work more admired. Finally, by about the second century B.C. they aspired to nothing less than duplicating the most creative forms of human behavior with their self-propelled series of mechanical components. This idea of duplicating the powers of life by mechanisms

> must have reinforced the highly speculative thesis of the atomic theory that all existing things are complex mechanisms…
>
> (Brumbaugh 1966: 5)

Although Sylvia Berryman has contested and problematised the classical idea of the mechanical analogy (2009), it is undoubtedly the case that mechanics were largely modelled on organics. The extension of this idea that the organic universe itself was essentially 'machinic' becomes more powerful in early modern society – specifically with the refinement of the idea that the human is essentially a machine. The idea of the automata in Greek myth has a long history and the automata itself was something which was designed by its maker to serve and protect. In essence the machine was servile and at the bidding of its master. But as slaves are wont to do, the automata had enough independence and sentience, to try and escape from their master when they could:

> The wonderful statues attributed to Daedalus come from, and refer back to, a period halfway between the world of science and mechanics and its precursor, a world of magic. The truth about these statues, which "had to be kept chained, or they would run away," has been a real challenge to scholarly ingenuity. In addition to their propensity to run away, there was something remarkable about the eyes of the statues: either they moved or in some other way gave the impression that they could actually see. Unlike the general inventions the Minoans and Myceneans attributed to Daedalus, the story of the statues suggests some specific and remarkable innovations.
>
> (Brumbaugh 1966: 25–26)

The anthropic qualities and properties that some automata possessed confused the boundary between sentient beings and mechanics. The mythology of the sentient eye hinted that there was more to the making of automata than mechanics. The fact that the machines could rebel, flee and swarm and potentially both re-engineer and replicate themselves has become part of modern dystopian cultures. For Judith Donath, in her study of new digital social machines, one aspect is that 'A machine is inherently sterile, inanimate, automated, unthinking. The "social machine" under its sinister interpretation processes people for their data; it automates relationships, atrophying the human dimension' (Donath 2012: viii). But in so many ways those machines are not sterile, inanimate and unthinking even if they are part of often vast, automated systems. Marx, as we shall see, argued that what Donath calls the 'atrophying' of the human, is an essential feature of machines. The all-powerful 'animates' coordinated the organic, conscious appendages of the human to become part of the automated 'social brain'. The 'social brain' that produced the machines in the first place then subjected humans (and other species) to their rule. The machine is, in fact, the objectification and product of consciousness – but of a collective humanity rather than of the individual.

Gerard Raunig has examined the 'concatenations and connections, compositions and movements that constitute a machine' (Raunig 2010: 19). From elemental machines modernity has begun to develop new types of vast, often 'unstructured' forms of 'Abstract' machines. As Raunig notes:

> Abstract machines are things like this, which themselves have no form, are formless, amorphous, unformed. Yet their unformed-ness is not to be understood here as a lack, but rather as the ambivalent precondition for the emergence of fear as well as for the invention of new, terrifying forms of concatenation.
>
> (Raunig 2010: 117)

The concatenations of beings like automata and even prosthetic humans are in their very essence hybrid and syncretic. They often bring together, in one being, a binary formation. The vast 'concatenations' of new machines can be combinatory of literally thousands of parts and modes of composition and recomposition often with no 'centre', 'brain' or 'heart' to them. If we look at the vast social machines of the web not only are there multiple tendrils, but there is essentially no central body which syncretises with its other additional parts. Like humanity itself (if we can extend the mechanical analogy to humanity) the machine is a seething mass of elements with its own enclaves, provinces, masters and subjected humans and species similar to what Hardt and Negri have called the 'machinic logic of the multitude' (Hardt and Negri 2001: 369).

We will be returning to accounts of the rise of the machine in historical practice, anthropology and classical sociology and specifically to the account of the machine by Marx and Engels. It is valuable to begin to think about three of the early commentators on the development of the machine in the medieval period and on into early modernity. Max Weber often used machine analogies in his exploration of bureaucracy and the ultimate fate of humanity. Durkheim's understanding of the division of labour, nature and totemism were more opaque in that his direct encounters and observations of the machine were limited. His initial thoughts, however, are worth pursuing, particularly the Durkheimian influence on the *Annales* school of historiography and specifically on that great commentator on the medieval machine Marc Bloch (see Hudson 2016b for specific accounts of Bloch's historiographical practice).

Marc Bloch and the machine

In his work on medieval inventions Marc Bloch continued his demonstration of the importance of technological history to social forms and vice versa (1967a). One problem in technological history for Bloch was the perishing of artefacts and an inability thereby to present a full structural history of the development of tools. Decorative forms and aesthetics were able to give precise notations of periodisation, influence and development in locations such as cathedrals and churches, but the history of the vernacular artefacts of their masons and carpenters were more opaque. For Bloch:

> A millwheel, a plough, a horse-shoe, a rustic spinning-wheel on the other hand have no style. Moreover the long and peaceful use to which these instruments have usually been put is an additional obstacle to fixing their precise date of origin. Our civilisation is mortal, but it is not yet dead: the archaeologist has a perfect right to say that it has at present no ruins.
>
> (Bloch 1967a: 172)

The archaic 'ruination' of archaeological strata and sedimentation could potentially offer clues to the development of technics and indeed has done (see Hudson 2017), but often it is in the phenomenology between individual human makers and users that the clue to adaptation lies.

Bloch notes the adaptations as a consequence of 'hand and mind and eye' – that relationship between consciousness and the material practice of individual human beings in social formations which did ultimately lead to a whole occidental 'technical civilisation' (Bloch 1967a: 175–176). This cannot be reconstructed either through the analysis of the aesthetics of successive styles and models, neither through excavation nor the reconstruction of human-material practice. History can only offer clues to adaptation and development rather than any form of historical reconstruction. Bloch then moves away from looking at specific technical artefacts in order to examine some problems in the social formation which might offer clues to changes and adaptations in technics. These clues in the social formation indicate that technics and mechanics displaced the organic for specific social reasons – 'the disappearance of the cruellest form of human subjection' (1967a: 181). As Bloch notes in a powerful rendition in understanding the entanglement of social and technical forms:

> Mills driven by water or by wind; mills for corn-grinding, for tanning, for fulling; hydraulic saw-mills, blacksmith's drop-hammers; shoulder-collars, the shoeing of draught-animals, harnessing in line; even the invention of the spinning-wheel: all these represent progressive steps towards a more effective use of natural forces, animate or inanimate, and hence led to economies in human labour, or – what comes to much the same thing – a more productive return. Why was this? Perhaps because there were fewer men available; but most of all because the master had fewer slaves.
>
> (Bloch 1967a: 181–182)

In essence, the decline of slavery, for Bloch, was causal for the technical revolution (1967a: 182). The economies of human labour though, far from freeing humans from subjection to harsh labour and slavery, would eventually make appendages of them to the machine itself – whether mill or factory. It would, as we shall see, also have significance for the plebeian animal.

Bloch argued elsewhere that technical innovation and its rejection was a collective psychological phenomenon (1967b: 125). This was nowhere more apparent than in the social imaginaries of technological fear at the upheaval that the machines might bring (1967b: 129). The terror at the idea of technical adaptation was not new and the machine-breakers of early capitalism were continuing a long tradition of contestation with the machines that were being imposed upon them. The technical civilisation that emerged in the occidental world was, for Bloch, essentially imitative and adaptive of the technical innovations of other social forms – specifically the inventions of medieval Islamic societies (1967b: 134). The utilisation of these inventions was essentially because of the unique social conditions which brought the medieval period into being in the first place including the decline of slavery and the ossification of technical invention in the Roman period (because of its extensive slave networks). This led ultimately to the mechanised system of industrial capitalism which is itself 'positively haunted by the technical world' (1967b: 134).

If the 'motive-power' of a society is essentially individualised labour using small machines like tools for their own subsistence, or social labour of vast numbers of slaves where no technological innovation is necessary to make labour more effective, then what happens when other forms of 'motive-power' displace those different forms of labour? In effect the answer is both technical innovation *and* insurrection. Nowhere is this clearer than in the relationship of social formations to different methods and machines of 'milling'. The motive-power of labour and the handmill was displaced by the motive-power of water and the watermill (1967c: 141). Bloch argues that the advent of the watermill was essentially a process whereby the collective (in this sense, feudal authority) imposed its order upon individualised labour. But in actuality the technical innovations had been present for other a thousand years without being taken up in any collective and social form. The direct human interventions into nature that we first see with the Neolithic agricultural revolutions and then with the development of machinery in ancient societies did not develop into an economic system until the social collective was ready for it. As Bloch says:

> But in the course of all this progress towards relieving the animate world of physical effort, the history of which more or less summarises the essentials of technical evolution – iron replacing wood, coal supplanting charcoal, chemical dyes taking the place of cochineal and indigo – in the course of this increasingly direct control exercised by man over elemental natural forces, without recourse to the power of animals, the advances made shortly before the birth of Christ were, in one sense, the most decisive of all.
>
> (Bloch 1967c: 141)

The advent of the watermill itself in the Roman period represented in material form a technological revolution but it was not socially extended throughout the Mediterranean and European social formations (Bloch 1967c: 142). The natural context and its geographical and geological modes of possibility were of course already present in the forms of stream and river systems and eventually milling would imitate and become parasitic upon these natural formations. Human intervention would in turn divert the courses of rivers and transform the natural water-systems as the watermill appeared (Bloch 1967c: 142).

The general social unwillingness to utilise either the natural formations or the technical innovations meant that machinery itself was not the 'motive-power' of the technological revolution but the unique nature of the collective social forms. In societies that were unwilling to extend 'dominion' over nature we see a parallel and contestatory process – that of the 'dominion' over human 'nature' and labour. As Bloch says, the Roman lords 'had no reason to install expensive machinery when their markets and their very houses were overflowing with human cattle', (1967c: 146). When the technology began to be socially adopted in the medieval period, the war against the individualised 'vernacular' handmill prosecuted by the feudal authorities in order to control local economies and populations, there were 'insurrections on the part of housewives' (1967c: 157). This became a war, especially in England, between 'wind and water against human muscle' in which the feudal authorities and the watermill were for some centuries victorious (1967c: 156).

The handmill and other vernacular technologies were never entirely suppressed amongst the labouring populations. In areas where feudal authority could not reach, the insurrection against the machines would continue into the capitalist period where the watermill would be replaced by the steam-mill and successive technological adaptations. As Bloch says, there was both a war and a contiguousness between the rival systems – 'In short, when the iron and coal age opened, the ancient prehistoric tools had nowhere yielded altogether to the "engines" which for so many centuries had also relied upon the inanimate forces of wind and water', (1967c: 159). The 'ancestral ways' (Bloch 1967c: 160) of the archaic peasantry continued into industrialised social formations as did the medieval machine into modernity with some commentators indicating that one route away from ecological catastrophe could be the reassertion of the more 'organic' (wood, water, wind) technologies of the pre-modern period (Gimpel 1983, 1992).

Classical sociology and the machine

Max Weber's observations on the machine form part of his larger corpus on economic forms and the role of rationality. Like Bloch, he focuses initially on two different forms of machine. Bloch focuses on the vernacular individual machine and then on the socialised, collective machine which subjugates earlier human relationships. Weber concerns himself with contrast between the watermill as a specific 'apparatus' servile to the human and then with the further development of the collective machine which makes the human servile to it. As Weber notes:

> We think at once of the steam engine and the mechanization of work, but the machine had its forerunner in what we call "apparatus"—labor appliances which had to be utilized in the same way as the machine but which as a rule were driven by water power. The distinction is that the apparatus works as the servant of the man while in modern machines the inverse relation holds.
>
> (Weber 2003: 302)

It is ironic then that the watermill which caused so many insurrections against it is contrasted positively with the next stage of machinic development – that of the factory apparatus. Not only is this less bound to the natural geologies of water courses all parts of nature are diverted towards its more effective functionality.

For Weber the relationship between nature and the industrial system was both critical and tenuous. He argued that the destruction of vast swathes of forest itself halted industrial development at certain points because there was not enough organic material to fuel its furnaces. The development of coal extraction, rather than forestry, attenuated the organic dependence on trees by shifting to another organic dependence on the coal strata (itself with catastrophic social and ecological consequences). The ongoing dependency of the iron foundries on forestry and in the flooding of mines was offset in the former by the use of coke and by the latter with the steam-engine which could clear the mines of water. The onslaught of nature against the machine was mitigated by further refinements in the machines themselves (Weber 2003: 302–305).

Another major problem for the development and refinement of the machine was the indispensability of labour as an appendage and caretaker for the machine. Even if the development of the machine, as Marc Bloch has said, either released labour or operated in spaces where labour was scarce (after the end of slavery) the continual development of the technological forces displaced large collectives of 'handworkers' in favour of smaller numbers of workers to supervise the machines which had taken over the labour of human beings (Weber 2003: 306). The development of the new labour forces contiguous to and dependent upon the machines directed labour in new technical, organisational and psychological directions. These were worker-militants subjected to the laws of armies – armies which were directed by the machine and its modes of coordination:

> No special proof is necessary to show that military discipline is the ideal model for the modern capitalist factory, as it was for the ancient plantation. However, organizational discipline in the factory has a completely rational basis. With the help of suitable methods of measurement, the optimum profitability of the individual worker is calculated like that of any material means of production. On this basis, the American system of "scientific management" triumphantly proceeds with its rational

> conditioning and training of work performances, thus drawing the ultimate conclusions from the mechanization and discipline of the plant. The psycho-physical apparatus of man is completely adjusted to the demands of the outer world, the tools, the machines—in short, it is functionalized, and the individual is shorn of his natural rhythm as determined by his organism; in line with the demands of the work and procedure, he is attuned to a new rhythm through the functional specialization of muscles and through the creation of an optimal economy of physical effort.
>
> (Weber 1968: 1156)

Not only was the machine subsuming and subjecting natural organic forms to its will (and destroying them in its relentless search for fuel) it was changing the very nature of the organic species of the human. The 'rhythms' it inherited as part of its organic ancestry and lineage was being replaced by the 'rhythms' determined by the will and necessity of the machine. Rather than the body operating within and conditioned by nature its physical form was being determined by its role tending to the machine. The physical effort of being a human appendage would create an entirely different muscular tendency dictated by the aspirations of the machine itself. The 'order' of machine production was one which human beings were literally 'born into' its 'mechanism' and would continue to be so 'until the last ton of fossilized coal is burnt'. The 'iron cage' was a phenomenon born of the foundry and the subjection of humans to its will (Weber 1930: 181).

Durkheim's relationship to nature and machines is of an entirely different order to Weber's. In fact, Weber's 'machinic' analogies are replaced in Durkheim by the 'organic' analogy – that social formations are essentially like living bodies. Yet Andrew McKinnon (2014) has argued that Durkheim often used tropes of the 'electric' to organise social life as energy. In a remarkable passage from *The Elementary Forms of the Religious Life* (2001) Durkheim explores the relationship between the organic bodies of the human and the social body in electrical metaphors:

> When a force or a property seems to us to be an integral part, a constituent element, of the thing it informs, we have trouble imagining that it can detach itself and transport itself elsewhere. A body is defined by its mass and its atomic composition; and we do not imagine that it can communicate any of its distinctive features through contact. But on the other hand, if we are speaking of a force that has entered the body from the outside, something that is not attached to it, something alien to it, it is not unthinkable that it might escape. So the heat or electricity any object has received from an external source can be transmitted to the surrounding setting. The mind readily accepts the possibility of this transmission. The extreme case with which religious forces radiate and become diffused is therefore not in the least surprising if they are

generally conceived as external to the beings in which they reside. And that is indeed implicit in the theory we have proposed.

(Durkheim 2001: 240)

The idea of the presence within the individual body of this external 'electric force' from the social body is rich in implication. What is it about contact that means transmission and in which directions does transmission take place? The alien electrical power that transforms the individual 'atomic mass' of the individual entity is a process of diffusion and although profoundly different from both Marx and Weber's notion of the machine, this analogy also holds something which is of critical important to understanding the human–machine relation – that of the division of parts.

Each individual is comprised of a number of segments just as each social form is comprised of its own series of segments. The division in the social body is then imposed or transmitted to the individual body – dividing it, segmenting it, vivisecting it. This segmentation then reflects the exterior social form which created those caesurae. If then the causal structures of the machine are the dominant social and economic form in a society then the physical and the interior being of the individual is segmented into those parts useful for the machine and its modes of governance. In reality the machine becomes the new totemic expression of the social form, its various parts divided not into symbolic structures, but the actual machinic elements of the system imposed upon human beings.

The 'regular succession of phenomena' perceived by the individual is itself subjective argues Durkheim, and only becomes a fact when this regularity is social and collective. Causality then becomes the manner in which our 'empirical ascertainments' about our society and other human beings become understandable (Durkheim 1915: 368). The regularities and causalities that we observe in both nature and in our machines are the ways we understand ourselves, others and our social cosmologies. If we can extend this insight further in terms of our position and our modes of observation then Durkheim argues that our individual observation of regularity and similarity is one thing. The notion of the 'Genus' is something different:

> Genus is the external framework, whose contents include objects perceived as like one another. The contents cannot provide their own framework. They consist of vague and shifting images, the superimposition and partial fusion of a fixed number of individual images found to have common elements. By contrast, the framework is a definite form with firm boundaries that can be applied to an indefinite number of things, whether visible or not, whether actual or merely potential.
>
> (Durkheim 2001: 113)

The organisation and classification that we have as individual observers is different from the classifications, organisations and divisions that the social form imposes in *actuality*. The disjunction or conjunction between our modes of observance and the 'Genus' creates our social imaginaries over our social spaces including that of nature (2001: 339). This returns us to our human relationship to nature:

> But if the categories at first translate only social states, does it not follow that they can be applied to the rest of nature only as metaphors? If their sole purpose is to express social things, they could be extended to other areas only by convention. Insofar as they serve us for thinking about the physical or biological world, they would have only the value of artificial symbols, practically useful, perhaps, but with no connection to reality.
>
> (Durkheim 2001: 20)

The categories of our social states then sever our relation to nature and the physical and biological world because we can only approach them in the metaphors which are offered to us in our divided, vivisected social being. They are only able to be partially revealed by analogy. Actuality lies only in the social body of the collective and not in our own, partial beings. The vivisected social beings who are the appendages of the machines, in the factories, in the whaleships, in the industrial slaughterhouses have little capacity to understand nature except as metaphorical extensions of their own being and of their own vivisection. The social body of the ship as a whole 'wooden world' or of the industrial machine as a vast, abstract concatenation creates humans in its own image. The proletarianisation of labour in the face of the machine itself diminishes the species and the organic entities with which it reproduces and can only reproduce itself through food, fuel, wage. The suppression of nature and the subjugation of other species are the imposition of a social totality in which the very interior of human beings reflects the fractures and the horrors of the whole social formation but in micro-form.

Machine aesthetics

Marc Bloch had noted, in his work on medieval machines and specifically the mill, the specific *aesthetic* problems of understanding their development. 'Vernacular' machines were not only perishable but partly because of their ubiquity, their stylistic features can only give imprecise notions of their periodisation and their specific place within the structural development of that particular artefact. The aesthetics of machines are part of the wide-ranging discussion in later chapters but it is worth noting here some of the initial issues about technology and art, particularly in relation to the twentieth century where the 'aesthetic' ceases to be purely about decoration and becomes more attuned to technicity and functionality. The remarkable catalogue of 'Machine Art' from the New York *Museum of Modern Art* from 1934 at least partly initiated a new aesthetic understanding of machines by the European and American avant-garde (Museum of Modern Art/MOMA 1994).

Alfred J. Barr in his introduction to the catalogue notes that 'Machines are, visually speaking, a practical application of geometry. Forces which act in straight lines are changed in direction and degree by machines which are themselves formed of straight lines and curves' (MOMA 1994: 12). The specific beauty of machine art, for Barr, rested upon the regularities and repetition of its geometries and the rhythm of its movements (MOMA 1994: 12). Barr notes:

> Motion is an essential function of many machines and sometimes increases their aesthetic interest, principally through the addition of temporal rhythms, both of movement and sound. The pistons of a locomotive or the rising and falling frames of a mechanical loom illustrate the point.
>
> (MOMA 1994: 12)

The suppression of ornament and decoration then opens up the sensuous beauty not just of its geometries but of its materiality and the materials used for its production (MOMA 1994: 13). This leads to perhaps one of the most important statements about the relation between technics and the aesthetic form of the machines. Barr writes:

> The beauty of the machine art in so far as it is a mere by-product of function may seem a meagre and even trivial kind of beauty. But this is not necessarily so. The beauty of all natural objects is also a by-product – the helix of a snail's shell (and a steel coil), the graduated feathering of a bird's wing (and the leaves of a laminated spring), the rabbit's footprints in the snow (and the track of non-skid tires), the elegance of fruit (and of incandescent bulbs).
>
> (MOMA 1994: 15)

The aesthetic appreciation of the 'natural form' is replicated in the machines that are produced in the 'conquest of nature's hostile chaos… Not only must we bind Frankenstein – but we must make him beautiful' (MOMA 1994: 15).

This remarkable passage not only reveals the status of those who observe and admire machine aesthetics but also their understanding and even hostility to a nature of 'chaos'. If machine 'beauty' can rest upon functionality (as does the snail shell) and the order of the machine, why then is nature to be subdued as 'chaos' when its own forms are also functional and 'ordered'. Nature's order on the one hand is a function of human classification systems and taxonomies of 'ordering' and on the other, nature, over and above human beings, has its own analogical orders to the human world. The monstrous hybrid products of nature and machine collision – the 'Frankenstein' of our invention – should not be hideous. In fact its true beauty comes from its functionality on the one hand and on the other from its materials. In this sense the materials that made Frankenstein – limbs, hands, head – are themselves simply transposed products of natural forms rather than pure machinic inventions.

The aestheticisation of machines and technology has a long history and we will return to the aesthetics of locomotive cultures later but it is worth raising a series of issues about machine art and its relationship to nature. Often, rather than understanding the machine aesthetic in and of itself it is transformed into a social imaginary which has its origins and locations elsewhere. The important factory documentation of Lady Bell saw the factories and foundries (belonging to her husband) as akin to medieval castles – 'On either shore rise tall chimneys, great uncouth shapes of kilns and furnaces that appear through the smoke on a winter afternoon like turrets and pinnacles' (Bell 1985: 12). The development of Gothic municipal architecture was essentially an architecture of industrialisation and the demonstration of the ideological sanctity of the ironmasters, whalers and owners of the mines. The persistence of either organic forms or the replication of organic forms in the architecture of factories, ships and towns expressed not a functionality, but an aspiration to found a new civilisation comparable to the one which had been displaced. The rural homes of the gentry were being taken over by the new ironmasters and often employed the new faux-medievalist furniture makers and designers to enrich their estates (as did the Bells with Philip Webb, see Kirk 2005).

The emerging dissatisfaction with this syncretic medieval-industrial architecture and planning would begin to see the development of new avant-gardes of futurism which in their geometrical rejection of natural form would fully abandon what they saw as the traditions of the 'dead generations' and the 'poetry of the past' (in Marx's term, see Hudson 2000, 2002). The gradual distantiation from any reliance on 'nature' and 'natural form' would create new ways of thinking about machine art as pure functionality, in and of itself, rather than replicating the nature which it aspired to replace. The development of new forms of locomotive cultures would eventually discard decoration, unwarranted symbols, and anything that looked like a natural analogy. This would be at the same time as it became more attuned to what 'natural law' could offer it in terms of increased speed, reduced wind resistance, and fuel extracted from the demolished forests, coal strata and eventually from the electric power which in turn relied upon the former. In terms of art this process would reach its apotheosis in the development of a pure 'machine' art which would abandon any pretence of human creativity or authorship and adopt the geometries and fluidities of 'automation' in which, like the factory, the human was only an appendage to machine production.

The elimination of the human as both the subject and the author of art was also part of a process that tended to disinterest in other species and in non-human forms. The elimination of the human aesthetically was part of a process in which the animals of other species were eliminated physically.

Species, human 'nature' and machines

The extension of the human technosphere into nature meant irrevocable change on a vast scale for other species. Not only were other species used for fuel and food many of their habitats were transformed or eliminated. Certain types of favoured species accompanied human beings, others were used for labour and some were eliminated entirely. There was a diversion of the evolutionary journeys of many species as there had been for *Homo sapiens*. The increasing use of technology meant a transformation in the powers of the human just as much as it meant doom for certain species. The fragile co-existence between early humans and other species in which landscapes and geologies were left largely unmarked by human intervention ended with the extension of the Neolithic farming communities throughout the Middle East and into Europe. As they transformed 'their' world human beings felt some sense of mastery – confirmed later in the Early Modern period with the development of new scientific taxonomies (classificatory systems which also relegated other humans).

Recognition that these taxonomies were instruments of discursive and material power and increasingly aware of ecological catastrophe, new thinkers from the 1970s onwards began to question the validity of scales and hierarchies of nature. Often this meant new perceptions of humans as themselves animals or the anthropocentric extension of valued human characteristics to animals. It often questioned the inevitability of human domination and would eventually lead to programmes which would halt the boundless human domination of nature by 're-wilding' territories. Fundamentally, many of these projects would actually make problematic the very status of the idea of the 'human' concept itself. Giorgio Agamben is perhaps the foremost commentator on the human as an 'open' phenomenon:

> man is not a biologically defined species, nor is he a substance given once and for all; he is, rather, a field of dialectical tensions always already cut by internal caesurae that every time separate – at least virtually – "anthropophorous" animality and the humanity which takes bodily form in it. Man exists historically only in this tension; he can be human only to the degree that he transcends and transforms the anthropophorous animal which supports him, and only because, through the action of negation, he is capable of mastering and, eventually, destroying his own animality.
>
> (Agamben 2004: 12)

Roberto Esposito, in his reading of Heidegger, sees the extension of the idea of *humanitas* as an 'anthropocentric *ratio*' in German anthropology and its parallel in zoology as central to the 'racial hygiene' of the Nazi regime. Nazism essentially humanised German citizens at the cost of other 'bestial' sub-human beings discarded from the hierarchy of nature as unnatural forms.

Those humans discarded and subjected to violence and extermination were animalised – 'He was an animal-man: an animal with a human face or a man inhabited by an animal' (Esposito 2011: 81). Still, Esposito does not fully discard humanism and replace it by anti-humanism. He argues for the innate interpretative and material possibilities of the idea of the human and for its mutations (2011: 81–82). He argues that this set of possibilities, far from bestialising human beings, should in fact animalise them all the more effectively:

> This is valid both in relation to other living species, such as, first and foremost, animals, and in relation to non-organic forms and matter. The two things should not be separated: that is, the intra-specific relation between man and animal, dating back to the age of domestication, constituted the first anthropo-technical, i.e., self-transformative, segment of the entire process of hominization [*ominazione*]. For this, contrary to what is presupposed in the bestialisation of the enemy carried out by all – ancient and recent – racisms, engaged as they are in the symbolic fabrication of the man-animal, the relation to the zoosphere, that is, the animal world, has always amounted to a decisive advancement of human culture.
>
> (Esposito 2011: 84)

The question of human 'being' is of course decisive in the work of Heidegger himself. As early as 1919 Heidegger was already preparing *Being and Time* [*Sein und Zeit* 1926] (1962). From its opening pages, Heidegger questions the inability of previous philosophers to think about the question of 'being' (1962: 19). Although emerging in classical philosophy as a theme for investigation, researches into the concept were 'fragmentary and incipient' (1962: 21). The inquiry into and the 'interrogation' of being had to ask first what entities of being (*Seiendes* – that which is) disclose about themselves in the world and what is articulated in their being (1962: 27). Heidegger later discloses that the 'tool'/*das Zeug* and the idea of 'readiness-to-hand'/*Zuhandenheit* is part if the nature of being – even if that 'readiness' is not something that we can conceptualise (1962: 97–98 and see also Harman 2010: 18). New 'multi-species' approaches have to understand the concept of the human 'tool' as part of thinking about the human past but also human futures and possibilities (Agamben 2004, Haraway 2008, Hartigan 2015).

Reflections on the 'tool' and 'technics' must address the sustained and multiple work of Bernard Stiegler. Stiegler has attempted to uncover human 'nature' by dissolving the human into a technical formation. The cultural revolution of the *Homo sapiens* and its intervention into nature in prehistory defines any 'essence' that the human might have in its 'being'. Stiegler's multi-volume work (1998, 2008, 2010) on the relationship between technics and human history is an attempt to describe a fully technical being in evolution:

> The problem arising here is that the evolution of this essentially technical being that the human is exceeds the biological, although this dimension is

> an essential part of the technical phenomenon itself, something like its enigma. The evolution of the "prosthesis", not itself living, by which the human is nonetheless defined as a living being, constitutes the reality of the human's evolution, as if, with it, the history of life were to continue by means other than life: this is the paradox of a living being characterized in its forms of life by the nonliving—or by the traces that its life leaves in the nonliving.
>
> (Stiegler 1998: 50)

The biological organism and the prosthesis, the living and the non-living – each defines what the human became and continues to be. Relying on the work of Leroi-Gourhan (see Audouze 2002) Stiegler uses the metaphor of the 'passage' – a passage to the human which took place in prehistory. If the human had any kind of 'birth' it was in the early Palaeolithic adoption of tools. This is particularly important if philosophy is displacing the human or thinking about the end of the human (Stiegler 1998: 135). Human 'technical being' is what Stiegler calls a 'complex of heterogeneous forces' (1998: 2). And these forces are essentially material ones or interactions between biology and its technics. For Stiegler the development of the human cortical structures are enmeshed with the machine:

> One must first ask what mirage of the cortex is experienced [*s'éprouve*], as pathbreaking, in the hardness of flint; what plasticity of gray matter corresponds to the flake of mineral matter; what proto-stage of the mirror is thus installed. One must then ask what the closure of the cortical evolution of the human implies from the vantage of a general history of life, the closure of the cortical evolution of the human, and therefore the pursuit of the evolution of the living by other means than life—which is what the history of technics consists in, from the first flaked pebbles to today, a history that is also the history of humanity—a statement that will lead us to the unusual concept of "epiphylogenesis".
>
> (Stiegler 1998: 135)

Not only does this recall the interactions and relations between the organic and the machinic snail shell and steel coil that Barr noted above, it also returns us to the absence of the cortex in animals and whether this marks a decisive break between the animal and the human. For Philip Low and the *Cambridge Declaration on Consciousness* (2012) the absence of certain types of neurological structures in animals which are present in humans makes little difference because affective states are comparable – we are all animals and have similar forms of intentionality. What marks the break between the animal and human for Stiegler is its technicity born in the Palaeolithic tool revolution.

Pieter Lemmens, as part of an interview some years ago with Bernard Stiegler, has argued that the central theme of the latter's work is that paradoxically the human has no 'origin' because it is itself an 'accidental being' without any qualities except those that technics imposes upon it. The tool and the machine are in fact what define humanity and are 'co-extensive' with the nature of humans (2011: 33). Lemmens indicates Stiegler's recent indications that one of the significant current threats to human beings, in terms of both psyche and collective, is the new industrialisation of memory and cognition in the 'mnemotechnologies' of writing, printing and digitalisation (Lemmens 2011: 33–34). For Stiegler himself the schism between the human and the technical and the organic and machinic is artificial. Technicity has collaborated with the human since its beginnings (Lemmens 2011: 35) but the colonisation of 'memory' by industrial capitalism creates new dangers for our 'technical beings' specifically in what Stiegler has called the 'third memory'. As Lemmens notes in his commentary to Stiegler:

> For Stiegler, technology has to be understood first of all in terms of a memory, i.e., as a supplemental memory system unique to humans. Besides an individual neural and a specific (species) genetic memory, common to all sexualized living organisms, humans possess a third, technological memory or inheritance system which has enabled the transmission of individual experience over the generations, i.e., the possibility of individual experience becoming available for the species at large thanks to its inscription in technical artefacts (opening the historical mode of being Heidegger called 'existence'). Humanity has evolved on the basis of this technological inheritance system. This means that human evolution cannot be understood anymore in Darwinist terms, since Darwinian evolution presupposes that individual experience – 'acquired characteristics' – cannot be transmitted to the species.
>
> (Lemmens 2011: 41n)

'Being' then is inseparable from memory. If memory is neural, collective and technical and the mnemonic experience of the neural can then be 'transcribed' through the artefacts of writing, printing and digitalisation, then this has a series of effects. The acquisition of the neural and species memory by the technical means the extension and extra-territorialisation of highly specific memory forms away from their originary locations in cognition and individual memory and into globalised forms of inscription. This means that the provenance of individual memory becomes problematic and control of the uses to which memory is put becomes impossible. But it also offers possibilities. Not only does that technical inscription dislocate memory away from its specific origin, in this or that memory or text, it allows for the proliferation of multiple memories and sources that the technic humanity can then use as a mode of reflection upon its history. Its inscriptions in the sedimented archaeological records can be recovered and understood, the impact of humanity's interventions into the coal strata can be addressed, the 'human movement' can use those inscriptions to understand the havoc that it has wrought upon species. The texts of memory can provide globalised resources to think about the futures of human–animal–nature relations and avert us and them from catastrophe.

The catastrophe of species-destruction is already anticipated in the early writings of Marx himself. In his economic and philosophic manuscripts of 1844 and some years before the emergence of a mature materialist conception of history, Marx begins to delineate the relationship of what he calls 'species being' to 'social being'. In some ways this begins the rift between the metabolisms of humans and other species that we will encounter later – but formulated at the very beginnings of the Marxist project. On the one hand species being is the notation of the biological collective of humans and their entire material manifestation as a species over and above their 'subjective individuality' (1975: 327). But species being is also the manifestation within the individual of the whole of universal humanity (1975: 327). But the human is also a species-being because she recognises both their own species and that other species as their object of analysis and of physical practice (1975: 327). This is entirely different to other nonhuman species who are not conscious of anything beyond their biological 'life activity' with which they are one and inseparable (1975: 328). For Marx the technical domination of nature, of other species and of their own is central to the biological sentient 'species being' of the human collective. As Marx notes:

> It is therefore in his fashioning of the objective that man really proves himself to be a species-being. Such production is his active species-life. Through it nature appears as his work and his reality. The object of labour is therefore the objectification of the species-life of man: for man reproduces himself not only intellectually, in his consciousness, but actively and actually, and he can therefore contemplate himself in a world he himself has created.
>
> (Marx 1975: 329)

The world is the object of the human, it is the laboratory of practice and experimentation. These practices and experiments inscribed upon nature indelibly mark the earth as humans biologically reproduce themselves. The world is in fact a mirror for human beings. It recognises itself only in its practice upon the world. Further, those inscriptions are the memories of both the world in and of itself and of a history of human practice upon it. Production and reproduction thereby creates the social memory of the entire species as the memory of extracting, making and consuming. Yet production would ultimately lead to a form of estrangement, of separation and alienation from the object. For Marx:

> In tearing away the object of his production from man, estranged labour therefore tears away from him his species-life, his true species-objectivity, and transforms his advantage over animals into the disadvantage that his inorganic body, nature, is taken from him.
>
> (Marx 1975: 329)

The memory of practice and production, entwined as it is with the machines that humans use to inscribe the earth, will ultimately subordinate human labour into the sentient cogs of a productive, technical system. Indeed the memory and the sentience of human beings will condemn them to be the last organic remnants within a vast apparatus of machinery that they can no longer dominate or control.

In the next two chapters we look at respectively the notion of the industrial factory and that of the machinic whale-ship. Marx and Engels and their specific observations of the machine and Scoresby and Melville's ethnography of Cetaceans are specific inscriptions, made possible by technicity, which warn us against the fate that might ultimately subsume humanity as humanity subsumed and consumed others who did not share their species being.

2 Marx and the machines

In the second chapter, we look specifically at Marx and machines. In some ways, the most remarkable passages of Marx lie in the marginalia of his writings and in that of Engels. We examine specifically Marx's notebooks from the winter of 1857–1858 collectively known as the *Grundrisse*. Within this work there is a 'fragment' on machines but also the development of a concept of 'metabolism' which has become central to new thinking about ecology – specifically in the social theory of John Bellamy Foster and his collaborators. This chapter extends that thinking about 'metabolic rifts' between humanity and nature in new directions.

Marxism and nature

Inscribed into the origins of historical materialism and the Marxist conception of history was a double irony; first, that historical materialism is first and foremost the study of human 'natural' forms (species being) and second, that this natural form of the human has a will to completely dominate nature and ultimately forge a new society which manipulates nature in order to allow the natural form to thrive. The emergence of historical materialism was premised, in the manuscript of *The German Ideology* of 1846, on the organic necessity of the human body to eat, drink, labour and reproduce itself materially through production (with the manuscript itself left to the 'gnawing criticism of the mice'). Ellen Churchill Semple has provided a romantic description of this 'natural' human being:

> Man is a product of the earth's surface. This means not merely that he is a child of the earth, dust of her dust; but that the earth has mothered him, fed him, set him tasks, directed his thoughts, confronted him with difficulties that have strengthened his body and sharpened his wits, given him his problems of navigation and irrigation, and at the same time whispered hints for their solution. She has entered into his bone and tissue, into his mind and soul.
>
> (Churchill Semple 1911: 1)

Confronted by nature and its resources and looking to materially reproduce itself, the human collectives of tribe, language and custom (as Marx says in the *Grundrisse*) make and re-make themselves as they develop the productive forces of their societies. The emergence of class societies segment and 'vivisect' humanity even further. Ultimately the contestation between proletarian and master takes its penultimate social form in capitalism culminating in a transition to a new form of society or to the 'common ruin of the contending classes'. Further to the separation of human beings, the distinction between nature and the social has been the most sustained theme in the Marxist understanding of ecology. What is perceived as the nature/society dualism has obsessed commentators since the emergence of Schmidt's critical theory on nature in Marx (1971) particularly with regard to the political programme on the environment that might emerge from Marx's limited theorisation of nature (Benton 1989, Burkett 1996, O'Connor 1989). This work was reviewed in the early theoretical work by Noel Castree (1998, 2000b, 2002, 2005) who evades the dualism of naturalism and social constructionism by postulating, in his reading of the work of Neil Smith, the idea of the 'production of nature' (2000b).

Castree argues that Marxism has to stand both against bourgeois 'techno-centrism' which offers the amelioration of environmental problems and radical 'ecocentrism' which aspires to a reversion to pre-industrial or non-industrial modes of production (and by extension social forms contiguous to those modes of production). What Castree calls these 'antinomies' can be overcome by rethinking the abstract Marxist understanding of 'nature' as a natural or social production (2000b: 6). As Castree argues:

> ...the crux of the issue still remains: how to fashion a Marxist theory of nature which avoids the Scylla of a residual naturalism and the Charybdis of an anti-naturalistic social constructionism. Since my intention is not only to undertake a critique, the third aim of this essay is to offer a solution of sorts: namely, a little known reading of the Marx-nature-capitalism triptych which phrases its claims in terms of the *production of nature.* At first sight, this phrase seems only too redolent of the social constructionism to which I am objecting.
>
> (Castree 2000b)

By way of reading Latour on the 'modern', the notion of the 'production of nature' overcomes the separation between the social and the natural and the human and non-human (2000b: 12). This is possible by means of distinguishing a first and a second nature by anthropomorphically opposing the 'idea of an independent, non-social nature' (2000b: 25). As Castree proposes:

> I want to suggest that this non-dualistic conception allows Marxists to adhere to three things simultaneously: first, an ontology which, while it denies a separation between capitalism and nature, nonetheless refuses to

> elide one with the other; second, a supple, non-determinist theory which, in explanatory terms, accords power and agency to both capital and nature; and third, a normative perspective which criticises the ecological impacts of capitalism on historically- and place-specific grounds without reverting to a politics of nature in or for itself. These three things, it seems to me, are the rewards to be had once one rejects the antinomian thinking informing previous work on Marxism and nature.
>
> (Castree 2000b: 27)

This supports new forms of 'Eco-Marxism' and programmes to transform ecologies without reverting either to the concept of nature, to the actuality of nature or to 'natural' or pre-capitalist 'natural' modes of production. As Castree notes – 'the production of nature is a continuous *process* in which nature and capital *co-constitute one another in temporally and geographically varied and contingent ways* (2000b: 27–28). The idea of the 'co-constitution' of nature and capital is, however it avers it is not (and this is because Castree does not intend to provide any reading of Marx himself), essentially another anthropocentric 'construction' and social 'production' of nature. There are other approaches – specifically in terms of the relation between machines and humans that can offer an alternative understanding of nature as an entity in and of itself as an 'actual' set of phenomena and materialities.

Joel Kovel, in his analysis of capital as the 'enemy of nature', turns to that text that is most revealing of Marx's conception of nature – the notebooks of the *Grundrisse* (2002). The constant self-expansion of the human at great cost to nature is central to the argument that Kovel makes about the nature–human dichotomy:

> The depth of Marx's insight should be appreciated: capital is quantitative in its core, and imposes the regime of quantity upon the world: this is a 'necessity' for capital. But capital is equivalently *intolerant* of necessity; it constantly seeks to go beyond the limits that it itself has imposed, and so can neither rest nor find equilibrium: it is irredeemably self-contradictory. Every quantitative increase becomes a new boundary, which is immediately transformed into a new barrier. The boundary/barrier ensemble then becomes the site of new value and the potential for new capital formation, which then becomes another boundary/barrier, and so forth and on into infinity – at least in the logical schema of capital.
>
> (Kovel 2002: 41–42)

The initial boundlessness of nature and the plenitude of natural resources offers at first a co-habitation between humans and nature and then with class society and capitalism the radical intervention into, and exploitation of, nature's materiality. The co-evolution of humanity with other species ends with the domination of those species by humans who are first and foremost 'trans-forming forms' (2002: 99). The development of human socialities creates what Kovel calls a 'super-body', an organic analogy for the forces and relations of production. For Kovel:

> The boundary of the super-body with pre-existing nature is made by means of *technology*. Tools are extensions of the body as well as transfer points of the body into material nature, and of nature into the body. Technology is always socially determined and the bearer of meanings constructed through language. It is not a collection of tools but a fabric of social relations, certain threads of which are nature transformed into tools.
>
> (Kovel 2002: 101)

If technology is always socially determined then it is also the case that sociality is technically determined. In fact, as we have seen with Stiegler the human being is nothing except a being of technics (1998). The distinction and discrimination between natural and mechanical form is nowhere better expressed that in *Capital* itself and the preparatory work of the *Grundrisse*. The expropriation of labour from 'the soil' and the emergence of vagabondage, specifically during the Henrician period of the sixteenth century created the 'discipline necessary for the wage system' (Marx 1976: 761). The relation of labour to the 'earth' and the soil (and we will see the importance of this when we examine metabolism) is that the earth is both a laboratory for experimentation into production and reproduction and a repository of materials for that production (Marx 1964: 81).

What Marx calls the 'flight of the serfs' is itself the estrangement of the new 'vagabonds' from their fields (Marx 1964: 128) and the separation of the human from 'nature' itself which only becomes present to them in the form of the raw materials of the machine production. The independent activity of labour is replaced by the actuality of labour as 'appendage' to the machine. The centralisation of manufacture is at once the centralisation of capital, of populations and of raw materials. The division of labour, the application of steam, and the application of machinery, Marx argues, have put the world 'out of joint' creating unnatural forms out of the forces of 'Nature' (1975: 325). Many years later Theodor Adorno would argue that 'Technology is making gestures precise and brutal, and with them men. It expels from movements all hesitation, deliberation, civility. It subjects them to the implacable, as it were ahistorical demands of objects' (Adorno 1974: 19). The diminishing of nature then and the destruction of habitats and species to drive the machine is not the consequence of the human being per se, but a particular segment of that species which compels discipline and servitude to its own human mastery just as it subordinates everything else. For Marx:

> What we have here as an essential relation of appropriation is the *relationship of domination*. Appropriation can create no such relation to animals, the soil, etc., even though the animal serves its master. The appropriation

> of another's *will* is presupposed in the relationship of domination. Beings without will, like animals, may indeed render services, but their owner is not thereby *lord and master*. However, what we see here is, how the *relations of domination and servitude* also enter into this formula of the appropriation of the instruments of production; and they constitute a necessary ferment of the development and decay of all primitive relations of property and production. At the same time they express their limitations. To be sure, they are also reproduced in capital, though in an indirect (mediated) form, and hence they also constitute a ferment in its dissolution, and are the emblems of its limitations.
>
> (Marx 1964: 102–103)

The severing of the social and natural relation of the human to the soil is the appropriation of human will in social relationships of compulsion. In order to 'naturally' reproduce its own organic being labour is compelled to enter those relationships of domination and servitude and to service the machine as an instrument of production. The 'Beings without will' are appropriated, dominated, enslaved and destroyed, their habitats invaded, their flesh used for fuel. But for Marx the separation between the human and the non-human is expressed in consciousness and sentience – a concept made even more problematic over recent years by the *Cambridge Declaration*.

The Marxist understanding of nature then rests upon a series of suppositions – that it is possible to separate nature and the social, that nature can be produced socially, that the human and its technics are separate but in co-evolution. And commentators on Marx have postulated new ways that we can approach the challenges of ecological crisis without resorting either to the anti-humanism of a radical 'ecocriticism' while also surrendering the idea of the human as expansive possibility standing *against* nature and for the boundless increase in human technology – that we can observe, as we will see, in some variants of Marxism. Yet the relationship between species, nature, humans and machines has to offer a closer reading of Marx's key work on machines – that of the *Grundrisse*. Indeed, it is in the reading of the *Grundrisse* by commentators such as Foster that a new explication of nature and society as metabolism can begin to emerge.

Reading the *Grundrisse*

Adorno and Horkheimer, have argued that human thought in the twentieth century began to model itself not upon the aspiration to emancipate but the will to enslave. This was specifically rooted in the replication, within human thinking, of 'machinic' compulsions. They argue that 'Thinking objectifies itself to become an automatic, self-activating process; an impersonation of the machine that it produces itself so that ultimately the machine can replace it' (1997: 25). Not only is the machine about the production of the 'objective' world out there, but of the 'subjective' world within.

Antonio Negri has presented a specific reading of the *Grundrisse* that stresses the antagonism between the human and the machine and the emergence of different types of social subjectivities. Negri calls the *Grundrisse* an 'open work' and like Martin Nicolaus perceives it as problematic due to its incompleteness and fragmentation (1991: 1–4). Negri argues that what has come to be known as the *Fragment on Machines* was written at the end of February 1858 'at the peak of Marx's theoretical tension in the *Grundrisse*' (1991: 139). The development of antagonistic 'worker-subjectivities' around the machine that Negri elaborates from Marx he sees as rupturing the relation of capital from within the relation of capital itself. Interactions around the machine break capital 'from the inside' (1991: 149). The social individual becomes no longer an appendage to the machine but is itself what Negri calls 'multilaterality' (1991: 149). Ultimately, and sustaining this rupture in the machinic networks of capital, the social subjectivity of labour manifests itself '*In the communist revolution, the individual is social.* Social but concrete, he is exaltation and overdetermination, expansion of enjoyment, founder of that expansion', (1991: 146). The 'self-valorisation' of those multilateral subjectivities is synonymous with 'liberty' in the pages of the *Grundrisse* (1991: 150). The adversarial, antagonistic subjectivity itself as a social phenomenon is the very mode of transition between capital and the new forms of society which are in birth (1991: 152, 167). Harry Cleaver in his political reading of *Capital* and the *Grundrisse* has argued that orthodox understandings of *Capital* as a guide to the dynamics of capitalist exploitation from above are problematic. In its stead, a political reading focuses on the 'initiative of those resisting and attacking capital' (2000: 11). The focus on the industrial factory has to be generously extended out as a concept to the 'social factory' of wage labour, paid and unpaid, outside of the factory itself (2000: 70–71).

Casarino has seen both the *Grundrisse* and *Moby-Dick* (to which we turn in the next chapter) as expressing the dialectical tensions in the bodies of labour itself – of the 'indomitable excess of the dialectic that both Marx and Melville see incarnated in the living flesh of labor' (2002: xx–xxi). The entanglement between the human and the machine finds no clearer expression than in the *Grundrisse*; nowhere is the horror of the entanglement of the machine with other species better expressed than in Melville's work. But the *Grundrisse* is by far the more mysterious work. Due to its fragmented and opaque nature, and the fact of it remaining unpublished (if not actively suppressed) until the 1950s, it remains enigmatic and difficult. The *Grundrisse* is essentially the conglomeration of seven notebooks written by Marx during the winter of 1857–1858 and consists of two core chapters – 'On Money' and 'On Capital' with appendices. As its foremost commentator, Martin Nicolaus, has noted as did Negri later, there are 'missing elements of grammar, if difficult, sometimes awkward, obscure and even altogether inaccurate formulations, endless sentences and paragraphs, irritating digressions and reiterations', (1973: 24). As Nicolaus has said – it is a hazardous text from which to extract meaning (1973: 25). But a close textual reading of the *Grundrisse* can actually transform our understanding of the meaning of labour and production, of Marxism and of nature itself. The specific 'fragment' on machines is itself probably the most original and decisive description of the materiality that so marks historical materialism.

David McLellan, in his early reading of the *Grundrisse* and pre-dating the new translation by Nicolaus (McLellan 1971), sees the work as central to understanding the antagonism between labour and capital. For McLellan, specifically examining the machines fragment: 'Marx extrapolates tendencies inherent in his own time and foresees automation and technology as major factors in man's alienation; but he sees them equally as factors bringing about the collapse of capitalism and the transition to a future communist society', (1971: 154). McLellan also postulates the essential unity between the *Grundrisse* and the early works of Marx – specifically on 'species-being' in the 1844 Manuscripts. It is evidently a continued meditation upon and mediation of the themes of the early manuscripts (1971: 22). Seeing the *Grundrisse* as probably the only work of Marx to synthesise his central themes, McLellan offers another insight to their fragmented meanings and in fact sees the notebooks as the unified rendition of Marxism per se: 'In a sense, none of Marx's works is complete, but the completest of them is the *Grundrisse*' (1971: 25). This unified conception of Marxism finds its highest expression in the repeated motifs of production and reproduction in the *Grundrisse* not just of material products, but labour, and indeed species.

Production is itself a function of *appetite* first, of consumption. Human beings reproduce themselves by materially sustaining themselves with food and the accoutrements of life such as shelter and clothing. Beyond a certain point accumulation becomes more than subsistence and property emerges. Consumption in capitalism is on the one hand about sustaining and reproducing the material entities of the human body and cultures, but also about extending those appetites. The metropolis of the nineteenth century accrued to itself and its markets the fruits and flesh of the world and was part of vast networks of extraction, transportation and accumulation. Subjugating nature, its nautical and locomotive machines connected empires and colonies with its centralised, mercantile 'social body' in London, Manchester, New York, Paris and Berlin. The congealed evolutionary accumulation of human labour and machines was essential to the development of capitalism and that breach between labour and soil created a division of human beings from their hitherto methods of production. Production and consumption, for Marx, mediated one another (1973: 91). Capital was not just quantitative but organic and contained the congealed social memory of earlier human forms. For Marx:

> capital is – an instrument of production, for, in the broadest sense, every object, including those furnished purely by nature, e.g. a stone, must first be appropriated by some sort of activity before it can function as an instrument, as means of production. According to this, capital would

> have existed in all forms of society, and is something altogether unhistorical. Hence every limb of the body is capital, since each of them not only has to be developed through activity, labour, but also nourished, reproduced, in order to be active as an organ. The arm, and especially the hand, are then capital. Capital would only be a new name for a thing as old as the human race, since every form of labour, including the least developed, hunting, fishing, etc. presupposes that the product of prior labour is used as means for direct, living labour.
>
> (Marx 1973: 257)

The appropriation of materials and their transformation into prosthetic tools was indeed part of human society since the Palaeolithic era. The accumulation of fine movements over evolutionary generations was also the accumulation of capital – direct, living labour is simply the consequence of an archaic lineage of organic gestures and movements. These organic gestures and movement were themselves intentional movements to transform the world around the human and the collective and machinery is just a further extension of this process of production and reproduction (Marx 1973: 389). This in turn subordinates human beings as appendages to their own inventions. For Marx, capitalist production just extends the prosthetic process:

> The combination of this labour appears just as subservient to and led by an alien will and an alien intelligence – having its *animating unity* elsewhere – as its material unity appears subordinate to the *objective unity* of the *machinery*, of fixed capital, which, as *animated monster*, objectifies the scientific idea, and is in fact the coordinator, does not in any way relate to the individual worker as his instrument; but rather he himself exists as an animated individual punctuation mark, as its living isolated accessory.
>
> (Marx 1973: 470).

The negation of labour in the machine-oriented order is achieved by its subordination to the 'alien' will and intelligence of the *animated monster*. It is alien because its *animating unity* is elsewhere. The human appendage is an accessory that is neither coordinator nor instrument. What then is the role of organic, sentient labour within the machinic process and where is the elsewhere, the 'alien' location from which the machine has emerged and from which it gets its animating power?

If we can return again to *earth* and its relationship to labour and then, by way of this, return to a further elaboration of the machine. Marx argues that in pre-capitalist economic formations there is a distinctive relationship between labour and its appropriation of nature. If we can quote Marx at some length because this is the decisive statement on nature:

> (1) Appropriation not through labour, but presupposed to labour; appropriation of the natural conditions of labour, of the *earth* as the original

> instrument of labour as well as its workshop and repository of raw materials. The individual relates simply to the objective conditions of labour as being his; [relates] to them as the inorganic nature of his subjectivity, in which the latter realizes itself; the chief objective condition of labour does not itself appear as a *product* of labour, but is already there as *nature*; on one side the living individual, on the other the earth, as the objective condition of his reproduction; (2) but this *relation* to land and soil, to the earth, as the property of the labouring individual – who thus appears from the outset not merely as labouring individual, in this abstraction, but who has an *objective mode of existence* in his ownership of the land, and existence *presupposed* to his activity, and not merely as a result of it, a presupposition of his activity just like his skin, his sense organs, which of course he also reproduces and develops etc. in the life process, but which are nevertheless presuppositions of this process of his reproduction – is instantly mediated by the naturally arisen, spontaneous, more or less historically developed and modified presence of the individual as *member of a commune.*
>
> (Marx 1973: 485)

The 'social labour' of this formation is mediated for Marx through their occupation of the land. But 'individual labour' neither owns nor accumulates from the land as an isolated entity. This is made possible only through the commune or the tribe as a collective (and by extension the social classes) that direct its functioning – its social brain (1973: 485). As Marx notes in the above passage the workshop, laboratory and repository of nature presents itself to socialised labour as *earth*. This *earth* as the 'objective condition of his reproduction' is the basis for extraction and accumulation that are central to that reproduction (if only in terms of subsistence). The historical development of the *commune*, the earth, and the flesh of the organic body are all presuppositions of extraction and reproduction even when the flesh of the organic body precedes that body's activity and intervention into nature. Born into nature, as an animal, only the presence of the commune makes labour *social*.

The human organism is contrasted by Marx with the inorganic conditions that it reproduces itself from. In a key section of the *Grundrisse* he writes of the separation between them – particularly in its most extended form in the relation between wage labour and capital (1973: 489). This unity and separation of the organic and inorganic becomes of huge importance in the sense of 'metabolism' and the metabolic rift between humans and nature expressed by John Bellamy Foster as we shall see below. As labour is transformed by capital into an appendage of the machine and even further estranged from nature by the 'alien' monster of automation, the human, like other natural beings, ceases to be 'an accessory of the earth' (1973: 489). The rift between human and the *earth* of their emergence and previous productive forces becomes even more complete and the laboratory and materials of the soil become distanced from them. This returns us to the machine as an order of 'fixed capital'. For Marx:

> But, once adopted into the production process of capital, the means of labour passes through different metamorphoses, whose culmination is the *machine*, or rather, an *automatic system of machinery* (system of machinery: the *automatic* one is merely its most complete, most adequate form, and alone transforms machinery into a system), set in motion by an automaton, a moving power that moves itself; this automaton consisting of numerous mechanical and intellectual organs, so that the workers themselves are cast merely as its conscious linkages.
>
> (Marx 1973: 692)

The machine is not the individual worker's means of labour or her 'instrument' of production. The activity of the living sentient being around the machine and their interactions with it do not imprint upon the machine. Nothing is transmitted to the machine from labour. Indeed, it is the opposite – sentient labour 'merely transmits the machine's work, the machine's action, on to the raw material – supervises it and guards against interruptions' (1973: 692). The inorganic machine is master and virtuoso and consumes its materials and fuels as the organic worker consumes food (1973: 693). One element of this is intriguing and we will return to it in the next chapter. Marx makes a distinction between the organic and the inorganic but the fuel for the machines is often itself organic in origin even if now petrified and all life, but for its residual energy, extinguished from it. Coal, wood and even animal fuels congealed that residual energy from their earthly existence as living being that is transformed into the fuel for the machines. This would anticipate the extension of the use of human fuel in the industrialised death camps of the twentieth century and was part of huge accumulative processes in the Atlantic and Polar regions to source and accumulate whale oil just as much as it operated in the destruction of the European forests.

In any case, the conscious linkages of the sentient labour, guarding the machine against interruption also means something else for Marx. As we see above Marx sees the machine as a composition (literally a virtuoso) of mechanical and organic elements. The machine, although its automotive power lies 'elsewhere' and is 'alien' to the human workers is actually a syncretic, hybrid being comprised of both technics and biology. This is not to say that the human is becoming a machine or is analogical to a machine. Quite the reverse. The machine needs its sentient flesh as part of its operation – not just for fuel but for some form of 'conscious' operation. Why does the machine need this? And in what form does the antagonism between the machine and its human appendages appear? And who or what formed the machines in the first place and where did they come from?

Marx argues that the worker's activity is determined by the machine and not vice versa. The automaton is not driven or initially generated by the worker's consciousness or will. It acts upon the worker as the power of the machine itself (1973: 693). For Marx

> Labour appears, rather, merely as a conscious organ, scattered among the individual living workers at numerous points of the mechanised system; subsumed under the total process of the machinery itself, as itself only a link of the system, whose unity exists not in the living workers, but rather in the living (active) machinery, which confronts his individual, insignificant doings as a mighty organism.
>
> (Marx 1973: 693)

The 'ruling power' of the machine indeed cannot do without its active, living components, however petrified they might be. The social body and consciousness of the worker is entirely subsumed as part of the functioning of what then becomes this organism/inorganism – the hybrid of flesh and machine. Antagonism appears from within this machinic network, as Negri has argued, by the conflict between the automative will of the alien machine and the capacity of the worker for self-reproduction as a biological entity. The diminishment of all human capacities except those that do the will of the automaton limits and denigrates the capacity for reproduction. The remnants and *disjecta membra* of the human that still exist within the singular biological entity of the worker come into antagonism with the machine because of the very subordination *within* and *inside* the machine of the human flesh and consciousness.

So where does the machine itself come from? Not from individual labour but from the accumulation of social knowledge. This is knowledge which enhances the productive forces of an epoch which then returns to wield its power over the subsequent organic appendages of labour that find themselves within its technological system and rationale. The alien power actually only appears as an alien power – in fact it is the product of the 'social brain' or 'social body' of a society and its accumulated knowledges reappearing in monstrous form to the individual labour (Marx 1973: 694). The formation of 'social' labour within and inside the machine is necessary for the functioning of the automaton but also threatens its dissolution. As the machine is congealed 'fixed capital' accumulated by generations of investors and masters, so the machine is the congealed 'knowledge' of the social formation and its forebears. This brings us to Marx's most profoundly expressed elaboration of the 'social machine':

> Nature builds no machines, no locomotives, railways, electric telegraphs, self-acting mules etc. These are products of human industry; natural material transformed into organs of the human will over nature, or of human participation in nature. *They are organs of the human brain, created by the human hand*; the power of knowledge, objectified. The development of fixed capital indicates to what degree general social knowledge has become a *direct force of production*, and to what degree, hence, the

> conditions of the process of social life itself have come under the control of the general intellect and been transformed in accordance with it. To what degree the powers of social production have been produced, not only in the form of knowledge, but also as immediate organs of social practice, of the real life process.
>
> (Marx 1973: 706)

The 'social practice' of the machines are products of the 'human brain' but abstracted through collective, social life – the social brain and the social body. The will 'over' nature and 'participation' in nature have their emergence in the materials of nature that then subsumes and subjugates nature itself just as the social practice submerges those aspects of humans which are of no use to the machines. This returns us to what we might call the 'rift' in the human organism itself – the distinction between those things which are necessary and those which are unnecessary for the function of the automaton. If the machine is the product of human brain and human labour it has created a fracture in the worker. As the machine tries to refine the human muscle and consciousness to service itself as its organic appendage so those elements that are unnecessary become the sources of social antagonism and potentially, transition. Not that labour becomes 'machine-breakers' but that ultimately for Marx the machine is put into the service of human reproduction and not that of capital. Does this then transform the human intervention into nature and repair the breach between humans and the nature from which they emerged? In repairing themselves can we restore the metabolism to some kind of 'natural' functioning?

John Bellamy Foster and the metabolic rift

The idea of metabolism has its source in the *Grundrisse* and appears in the first volume of *Capital.* In this understanding labour mediates, across all social forms, the 'metabolism' between humans and nature (Marx 1976: 133). What Marx calls the *Stoffwechsel*, the metabolic interaction is itself a product of the human as a natural being and what it needs to reproduce its own organism from nature. The emergence of capitalist social relations and the machine disrupts the metabolic interaction between human and nature and particularly the human relationship to a diminished 'soil' or 'earth'. By socialising the means of production there might be for Marx, some sense of restoration of the 'earth' in order to further anthropocentrically enhance the powers of the 'human race' (Marx 1976: 637–638).

John Bellamy Foster has argued that the metabolic interaction has become a metabolic rift between human and non-human metabolism (1999). Further he argues that one central aspect of 'Eco-Marxist' political practice has to be the restoration of the metabolic unity between human and non-human. For Foster, in perhaps the most important critical analysis of Marx's 'ecological materialism':

> It was in *Capital* that Marx's materialist conception of nature became fully integrated with his materialist conception of history. In his developed political economy, as presented in *Capital*, Marx employed the concept of "metabolism" (*Stoffwechsel*) to define the labor process as "a process between man and nature, a process by which man, through his own actions, mediates, regulates and controls the metabolism between himself and nature". Yet an "irreparable rift" had emerged in this metabolism as a result of capitalist relations of production and the antagonistic separation of town and country. Hence under the society of associated producers it would be necessary to "govern the human metabolism with nature in a rational way", completely beyond the capabilities of bourgeois society.
>
> (Foster 2000: 141)

The metabolic rift between 'man and the earth' required restoration by the restitution of the soil and the return to the soil of the value that had been extracted from it. Foster explores the concept of metabolism further:

> The key conceptual category in Marx's theoretical analysis in this area is the concept of metabolism (*Stoffwechsel*). The German word "*Stoffwechsel*" directly sets out in its elements the notion of "material exchange" that underlies the notion of structured processes of biological growth and decay captured in the term "metabolism". In his definition of the labor process Marx made the concept of metabolism central to his entire system of analysis by rooting his understanding of the labor process upon it. Thus in his definition of the labor process in general (as opposed to its historically specific manifestations), Marx utilized the concept of metabolism to describe the human relation to nature through labor.
>
> (Foster 2000: 157)

The antagonism between the town and the country was also an antagonism between the metropolis and the periphery. The emergence of the peculiar accumulative form of colonialism saw their territories subject to extraction both in terms of natural resources and human labour leading to the global depletion of the 'soil' (2000: 164). So if 'metabolism' is the analogy by which we represent the 'human relation to nature' through labour and the 'metabolic rift' needs a form of political practice leading to the restoration of value to nature, then how can labour respond? For Foster, the disruption of colonialism and metropolitan supremacy by anti-colonial struggles is one part of this but also the whole project of Marxism needs to be reconfigured in order to reveal the essentially ecological core of Marx's thinking. The restoration of nature also means the restoration of labour to its own self-production rather than an appendage of machinery. Foster's 'ecological materialism' (2000: vii) is one which is situated in the techniques of Enlightenment science and in Darwin's materialist conception of nature (2000: 21). The restoration of an ecological Marx has to be situated within a materialist understanding of transmutation.

Yet the idea of the metabolic rift also reveals other problems. First, to return to the idea of the *Grundrisse* as a fragmented and mysterious text. Its refutation of meaning and interpretation is to make problematic the very reading of it, and of *Capital*, that Foster offers. Marx nowhere surrenders his obsession with the domination of nature for human purposes. In fact, a transition in a communist society serves the antagonistic social subjectivities of workers not the restoration of nature per se. Indeed, new forms of society would extend that domination, the rule of the machine and the further development of technology and not in the first place attempt to secure natural restoration. To see Marxism as a form of ecological politics is profoundly problematic.

Second, significantly larger parts of both the *Grundrisse* and *Capital* are concerned not with the minutiae of Marx's marginalia on nature (which is largely found in the sections on pre-capitalist economic formations) but with money, capital and fundamentally 'the machine'. It is the metabolic rift between the human and the machine that is far more significant than the metabolic rift between the human and nature. In a sense that rift, the accumulative and extractive practice of humans over nature, has been present since the birth of humanity and certainly since the development of Palaeolithic tool technologies and the Neolithic agricultural revolutions and migrations. If, as we have seen, machines are produced not only by the organic (in analogy and in actuality) social brain and social body, they are also meshes and collages of flesh and technics not just in terms of their sentient organic labour, but the materials they are made of and the fuels they consume. In actuality it is the restoration of the metabolic rift between the human and the machine that is of importance for Marx, not the restoration of some previous form of contract between the human and the amorphous non-human. This means that some kinds of 'Eco-Marxist' practices are in fact types of reversion to forms of society that are essentially pre-capitalist rather than post-capitalist. The defence of 'primitive' or 'feudal' social forms is preferable to the globalisation of technocratic capitalism and hence the development of 'anti-capitalist' political practice. If humans have always been, as Bernard Stiegler has argued, technical beings then any kind of restoration of lost and archaic natural and social forms is futile. The answer to our catastrophe may lie in the development of new social machines rather than in their supersession.

Engels, Manchester and machines

We will be developing the idea of the human 'encampments' of the city in a later chapter but Marx's concept of the machine owed not a little to the early ethnography of the industrial districts of London, Glasgow, and of Manchester in particular, of Friedrich Engels (see Marcus 1974). Engels himself later made clear that there were problematic formulations in this work – *The Condition of the Working Class in England* (1844) – specifically in that the materialist conception of history only fully emerges for Marx and Engels in the manuscripts of *The German Ideology* of 1847. He recognises that this early work on the Manchester factory system was part of the embryonic development of historical materialism but was essentially pre-materialist in origin and had only a vague understanding of capital, value and the role of labour (1969: 10).

Engels calls factory labour – 'this indirect trade in human flesh' (1969: 42). Interestingly, Engels here also uses the analogy of the English soil for the emergence of the classical proletariat:

> The history of the proletariat in England begins with the second half of the last century, with the invention of the steam-engine and of machinery for working cotton. These inventions gave rise, as is well known, to an industrial revolution, a revolution which altered the whole civil society; one, the historical importance of which is only now beginning to be recognised. England is the classic soil of this transformation, which was all the mightier, the more silently it proceeded; and England is, therefore, the classic land of its chief product also, the proletariat. Only in England can the proletariat be studied in all its relations and from all sides.
>
> (Engels 1969: 46)

Engels initially examines the proletarianisation of the 'handicraft' machines like Hargreaves' 'Spinning Jenny' which multiplied the production of cloth carrying as it did eighteen spindles operated by one labourer. Whereas previously there was a shortage of yarn there was now more yarn than could be woven by the labourers (1969: 47). This was the first powerful victory of the machines over its sentient appendages:

> With these inventions, since improved from year to year, the victory of machine-work over handwork in the chief branches of English industry was won; and the history of the latter from that time forward simply relates how the hand-workers have been driven by machinery from one position after another.
>
> (Engels 1969: 48)

Furnaces were built that were fifty times the size of the previous epoch and 'multitudes of iron artefacts replaced those of stone or wood (1969: 51). The centralisation of capital, labour and machinery in the great cities created the 'classical perfection' of the dual birth of industrial capitalism as a civilisation and its labouring masses:

> The Manchester Exchange is the thermometer for all the fluctuations of trade. The modern art of manufacture has reached its perfection in Manchester. In the cotton industry of South Lancashire, the application of the forces of Nature, the superseding of hand-labour by machinery (especially by the powerloom and the self-acting mule), and the division of labour, are seen at the highest point; and, if we recognise in these three

> elements that which is characteristic of modern manufacture, we must confess that the cotton industry has remained in advance of all other branches of industry from the beginning down to the present day. The effects of modern manufacture upon the working class must necessarily develop here most freely and perfectly, and the manufacturing proletariat present itself in its fullest classic perfection.
>
> (Engels 1969: 74–75)

This 'perfection' was of course present in a mitigated form to the labouring classes of the cities who lived, as in Manchester, amongst offal and debris and multitudes of pigs (1969: 80). This was a place of poverty, starvation, uncleanliness. Factory inspectors observed milk dripping from the breasts of women on the machinery whose children were drugged in the basements of tenements while their mother worked at the machine. Debris and carcases of animals were strewn across the passageways. The great human encampments were at once at the avant-garde of the new civilisation and also represented the most profoundly retrograde destruction of human labour, nature and animals. For Engels:

> The centralisation of population in great cities exercises of itself an unfavourable influence; the atmosphere of London can never be so pure, so rich in oxygen, as the air of the country; two and a half million pairs of lungs, two hundred and fifty thousand fires, crowded upon an area three to four miles square, consume an enormous amount of oxygen, which is replaced with difficulty, because the method of building cities in itself impedes ventilation. The carbonic acid gas, engendered by respiration and fire, remains in the streets by reason of its specific gravity, and the chief air current passes over the roofs of the city.
>
> (Engels 1969: 106)

In his mention of Carlyle's 'Proteus Steam Demon' (1969: 119) Engels begins to notate the enslavement of the organic entity of labour to the machine:

> In most branches the worker's activity is reduced to some paltry, purely mechanical manipulation, repeated minute after minute, unchanged year after year. How much human feeling, what abilities can a man retain in his thirtieth year, who has made needle points or filed toothed wheels twelve hours every day from his early childhood, living all the time under the conditions forced upon the English proletarian?
>
> (Engels 1969: 121).

The supersession of human labour by the machine (except as its sentient appendage) is, for Engels, intimately related to the musculature and the fine movements of those who intervene in the machines when it is interrupted. The gradual extension of steam and waterpower denigrates strength in favour of dexterity and unwittingly favours the entrance of women and children into the mills and the looms with every power, once having belonged to human labour as a natural and organic form, ceded to the machine (1969: 133).

Engels and nature

Friedrich Engels' late writings on nature have been the source of much controversy particularly in relation to Marx's work on nature and the idea of nature as 'dialectical'. In his textual critique of Eugen Dühring, Engels is concerned with developing two broad ideas. First, the relationship between humans and animals and second the distinction between human 'forces' and natural 'forces'. Freeing themselves from 'animality' allows the entrance of the human 'into history' (1934: 200). It also allows them to subdue and subjugate the forces of nature. Eschewing their status as 'half-animal' brutishness they become historical beings able to transform and conduct history according to their own social will. The emergence of class society and its productive forces created forms of sociality in which the domination of other human beings is itself a domination of nature and what can be extracted from it. Domination is by virtue of and '*through the intermediary of*, the domination over things' (1934: 208). As the historical being of the human becomes socially powerful the disentanglement from animality allows them to intervene in their social forms and into nature. For Engels: 'The forces operating in society work exactly like the forces operating in Nature: blindly, violently, destructively, so long as we do not understand them and fail to take them into account', (1934: 307).

The concept of humans as natural beings has to be displaced by a concept of them as social and historical beings in the domination that they can then achieve over nature. The forces of nature and the forces of the human are contestatory and adversarial: 'It is the difference between the destructive force of electricity in the lightning of a thunderstorm and the tamed electricity of the telegraph and the arc light; the difference between a conflagration and fire in the service of man', (1934: 307). The destruction of the rule and arbitrariness of nature is then, famously for Engels 'humanity's leap from the realm of necessity into the realm of freedom' (1934: 311–312). This 'world-emancipating act' is the task of labour in its confrontation with capital (1934: 312).

The fullest elaboration of this lies in Engels *Dialectics of Nature* whose outline plans give a guide to this unfinished work (1954: 27–28). Engels argues that natural science itself is a revolutionary practice (1941: 4). A large part of his analysis tends towards the critique of immutability that he finds in early modern mechanical conceptions of nature – particularly on the static classificatory system of naturalists (1941: 6). But by far the most powerful commentary that Engels offers lies in the 'nature' of human beings themselves specifically with regard to the hand and its dialectical relationship to the brain and by extension to machines and tool-making and tool-using technologies. Engels argues that natural forms, including human beings, arise through differentiation, division and distinction. The natural process itself biologically vivisects the human being into parts and by far the greatest and most powerful result of this is the *hand* as a tool (1934: 17–18). For Engels this is the source of the human subordination of and inscription upon (by hand) nature:

> Man alone has succeeded in impressing his stamp on nature, not only by shifting the plant and animal world from one place to another, but also by so altering the aspect and climate of his dwelling place, and even the plants and animals themselves, that the consequences of his activity can disappear only with the general extinction of the terrestrial globe. And he has accomplished this primarily and essentially by means of *the hand*. Even the steam engine, so far his most powerful tool for the transformation of nature, depends, because it is a tool, in the last resort on the hand.
>
> (Engels 1934: 18)

The subordination of nature means not just the subordination of the animal aspect of the human to its historical social brain, but the subjection of animals per se. The 'non-historic' nature of animals means that they are reserved by humans for subjection and relegation (and he would also do this for some 'non-historic peoples'). For Engels:

> With men we enter *history*. Animals also have a history, that of their derivation and gradual evolution to their present position. This history, however, is made for them, and in so far as they themselves take part in it, this occurs without their knowledge or desire.
>
> (Engels 1934: 18)

This history of the 'other' non-historic species can only be provided by humans. They can document the history, evolution, physiological development of animals. But the animals also have no history because history is an act of self-transformation, consciousness and sentience which, for Engels, they simply do not have. The more removed human beings become from their animality (we could say estranged or even alienated) 'the more they make their own history consciously' (1934: 18). Humans are the only species with a capacity to do this, have intentionality, make and re-make themselves, their world and other animal species. When they were literally of the soil and of subsistence humans had no conscious volition, acted as animals, fed and sustained themselves as animals. But the seizure of their history and their own natures meant the conquest and seizure of other natures and other species which were transformed and ultimately subjected to the productive aspirations of humans. Only humans can free themselves from animality: 'Man is the sole animal capable of working his way of the merely animal state – his normal state is one appropriate to his consciousness, *one to be created by himself*', (1934: 187). This accumulated autonomy and self-realisation is made possible only by production and reproduction. The self-production of human beings is the transition from collection (of natural resources and so on) to socially-formed productive processes. For Engels 'The most that the animal can achieve is to *collect*; man *produces*, he prepares the means of life in the widest sense of the words, which without him, nature would not have produced', (1934: 209).

This preparation of the 'means of life' has its foremost realisation in the article attached to the *Dialectics of Nature* in which Engels expresses 'The Part Played by Labour in the Transition from Ape to Man' (1934: 279). For Engels, it is the development of the hand and its tools in labour that was the propulsive mechanism for the development of the biological human. He argues that the hand is not just the organ most required for labour but is itself the evolutionary product of labour (1934: 281). The emergence of Simian proto-humans was intimately linked to the discovery and use of the properties of the natural objects that they encountered around them (1934: 282). The 'human conquest over nature' is a mastery of those objects and those animals. A mastery that we shall see may have its revengers. Engels notes that the subduing of nature is the essential property of the human and proto-human:

> In short, the animal merely *uses* external nature, and brings about changes in it simply by his presence; man by his changes makes it serve his ends, *masters* it. This is the final, essential distinction between man and other animals, and once again it is labour that brings about this distinction.
>
> (Engels 1934: 291)

The adversarial contest between humans and nature resulting in this conquest could have all kinds of unintended consequences by which animals and nature have their revenge upon human beings. The extension of the potato across Europe also meant the spread of scrofula (1934: 291–292). The extension of the humans and their machines has initiated an ecological catastrophe that could wipe out humanity. If a new ecological Marxism means anything it is saving humans from the consequences of their conquest, from their accumulations of knowledge and from their constant self-reproduction at the cost of habitats other than their own and species other than themselves.

The futures of nature

Some Marxists have seen the restitution of nature and that of labour as essentially the same humanising and naturalising process – that the struggle for nature and the emancipation of labour were part of the same process of transition to a new form of society not based on the exploitation of animals, habitats and the proletariat. The abdication of industrial technical civilisation is for many the aim of new types of Marxist 'ecocriticism' or at some level political practices which somehow mitigate the worst ravages of that civilisation. Yet still in Marxism, certainly much more than any other variant of classical sociology, there is a huge technocratic strand which sees the domination of natural forms as an essential parallel of the struggle for humanism. One might even talk about a protean humanism in which the human makes and remakes itself time and time again using the resources that it extracts from nature even as it extends the machines that had previously subordinated labour. In the closing pages of *Literature and Revolution*, Leon Trotsky formulates this technological revolution as part of his political programme against and over nature:

> Communist life will not be formed blindly, like coral islands, but will be built consciously, will be tested by thought, will be directed and corrected. Man, who will learn how to move rivers and mountains, how to build peoples' palaces on the peaks of Mont Blanc and at the bottom of the Atlantic, will not only be able to add to his own life richness, brilliancy and intensity, but also a dynamic quality of the highest degree. The shell of life will hardly have time to form before it will burst open again under the pressure of new technical and cultural inventions and achievements. Life in the future will not be monotonous...Even purely physiological life will become subject to collective experiments. The *human species*, the coagulated homo sapiens, will once more enter into a state of radical transformation...
>
> (Trotsky 1991: 282)

This remarkable passage, itself partly a reflection on the momentary incapacity of the Bolshevik revolution to initiate mass experimental building programmes, creates a distinction between the conscious, sensuous practice of human experimentation and the 'blindness' of nature.

Rather than an organic but non-sentient evolutionary process in which nature takes its directions in separation from human intervention, Trotsky proposes a process of experimentation, direction and correction *over* nature. Nature in this conception is not restored but entirely re-directed; the routes of rivers, the moving of mountains, the creation of cities at the bottom of the Atlantic. This is an entire disruption of nature including that of 'physiological life'. The collective experimentation will be one not just over inorganic nature but of organic species including human beings. The 'coagulated' species of the human will be utterly transformed time and time again. Trotsky goes further, in what amounts to his hyper-humanist and autocratic proposals for the species:

> Man will become immeasurably stronger, wiser and subtler; his body will become more harmonised, his movements more rhythmic, his voice more musical. The forms of life will become dynamically dramatic. The average human type will rise to the heights of an Aristotle, a Goethe, or a Marx. And above this ridge new peaks will rise.
>
> (Trotsky 1991: 284)

The physiological adaptation of human beings as collective experimentation and the interventions into nature that Trotsky proposes rest upon the idea, of which Trotsky was convinced, of the infinite malleability of nature and species. Further, it rests upon the notion of an earth that could infinitely sustain social experimentation and intervention into its mountains and oceans on the vastest of scales.

Yet this work was about the *aesthetics* of the machine age and the transitional social form that Trotsky was writing from within. The book itself was in essence, an argument against a purely proletarian culture (specifically in literature and architecture) and for an extension of the radical, technological humanism of the capitalist epoch and the Enlightenment. The obsession and experimentation with machines even extended to music and sound – exemplified by Avraamov's *Symphony of Sirens* (Smirnov 2013: 148–153). The emergence of musical machines, noise orchestras and symphonies written for factory machinery glorified the newly emerging industrial epoch (Smirnov 2013: 17). At the same time these experimentations in sound and the sonorities of machine cultures reflected the interventionist plans for the human species in the USSR. The preliminary programme that Alexei Gastev proposed for his *Institute for the Scientific Organization of Work and the Mechanization of Man*, included the development of a new 'human machine'. For Gastev: 'Anyone entering the front door of this institute as a normal living man, issues from the back door, after passing through countless laboratories, as a completely perfected working machine', (Smirnov 2013: 100). The metabolic rift, in this version of the future, is completely overcome – by the complete dissolution of nature, including human nature, into technology.

Of course, this was not a dystopian version of the future but an active living programme proposed as part of the wholesale social transformation of the early Bolshevik period. It foundered in civil war and the emergence of the technical apparatus of the Stalinist bureaucracy. Beginning with Lenin's proposal for the 'electrification' of the USSR it ended with the vast, nightmarish totalitarian cities of the Stalinist epoch – a peculiar and monstrous social formation itself a hybrid of space-age technology, despotism and social experimentation which for the exiled Trotsky differed from Hitler's fascism only in its more 'unmitigated savagery'. The physiological and natural 'collective' experimentation would continue in the USSR creating famine, destroying ecological habitats, compelling the mass migration of millions of people, and creating environmental pollution so profound that the effects on living tissue and genetics of humans and other species continues decades after its collapse. Commentators such as John Bellamy Foster and others have looked at a different route for an 'ecological' Marxist political practice and one which was not so enslaved to machines and the suppression of nature.

The suppression of other species and their subordination to socialised machines of execution, vivisection and transportation found its apotheosis in the vast 'Greenland' trade of the eighteenth and nineteenth centuries. At the very beginnings of industrial capitalist civilisation we can see the emergence of doctrines of natural domination based on three premises: that God had given us the flesh of animals for our food and fuel, that other species were part of a classificatory system of nature with humans at its pinnacle, and that 'industry' and 'invention' needed materials from nature for its effective and proper functioning as modes of accumulation. The effects of the 'Cetacean holocaust' are still with us today, without reparation and without restoration.

3 The Cetacean holocaust

In the third chapter we examine again the entanglement between species, nature and machines by uncovering the hidden history of a specific machinic industry – the historical 'animal genocide' of the 'Cetacean holocaust'. The rift between human and nature is nowhere more apparent than in the mass destruction of the whale populations of the North Atlantic in the seventeenth and eighteenth centuries by the whaling ships. Not only did this take place on a mass-industrial scale, the 'fruits' of the massacres were central to the development of the factory system.

At the same time as Enlightenment science was uncovering and classifying a whole taxonomic system of the Cetaceans, the wholesale slaughter of the whale populations created 'illuminants' for homes and factories, its oil greasing the machinery of the great manufactories, its labouring sailors becoming part of the first Atlantic proletariat. The entanglement of Cetacean metabolism with the human species and their machines was both catastrophic for whales and a symbol of a human dominion over nature which is now collapsing. Ernest Shackleton, in his memoirs of the *Endurance* expedition, the 'tale of the White Warfare of the South' (1998: vii) noted that the knowledge accumulated about whales and by whalers about the southern seas was central to the success of his mission to traverse the continent of Antarctica (1998: 1). Yet already, as the expedition journeyed into the pack ice, he noted that whales were taking refuge from their human persecutors in the open water within the pack ice. These ice-bounded oceanic territories provided sanctuary from the human world trying to capture and destroy them. Not only were his human encampments on the ice circled by whales, the documentation of the scale of whale hunting became a central finding of the expedition and indeed began a process of whale archiving which persists. By the time of the publication of this memoir the 'vigorous slaughter of whales' and their 'early commercial extinction' was to become part of the new impulse for conservation (if only to maintain a reasonable and continuous crop of whale products). These words themselves are from Robert S. Clarke's report on 'South Atlantic Whales and Whaling' from the expedition (Shackleton 1998: 370).

At the same time as the whale was being hunted to near-extinction new ways of interaction with whales were emerging. It is clear that these interactions are part of a long history of communication between whales and other species. Whales themselves are part of the creation myths of many peoples across the Earth (Billinghurst 2000: 40). The human obsession with the whale has often been seen as part of a complex and circular form of consumption. The myth of being swallowed by the whale, living within it, being expelled from its cavernous interior have long been part of human folklore. But humans also wanted to consume the creature, to subsume its physical flesh and its mental powers into their own bodies in order to reproduce themselves biologically and also to acquire the characteristics of the whale. This consumption would ultimately entail churches and boats being built according to the skeletal architecture of the whale. It would see the use of whalebone to create corsets that replicated the whale body. The oils of the vast machines of steam engines both in movement and inside the factory were extracted from the whale carcase. The bones of whales were used for human camps in the Arctic and for the gates of churches and fields. The near species-extinction of the whale would also be the first stage in the subjugation and subsumption of human bodies in the centuries to follow. The Cetacean holocaust was prefiguration and warning of the disasters to come.

Eternal Treblinka

Charles Patterson, in his work on the relation between animal subjugation and the Nazi holocaust, has utilised the idea of 'Eternal Treblinka' in order to elaborate the physical 'speciescide' of the animal world at the hands of human beings (Patterson 2002). Vasily Grossman, himself perhaps the most impressive commentator on both the Soviet and the Nazi death camps, which he and his family witnessed at first hand, has provided the central testimony of that first Treblinka (2011). Most remarkable is the testimony itself, that Grossman was in the first cohort of Soviet journalists to reach Treblinka. But even more powerful is the map of Treblinka that he drew in the first hours of reaching the camp. This map is witness to several things. First, that this place existed at all despite the attempts to erase it then and thereafter. Second, that in the location that the map represents hundreds of thousands of people had simply had their materiality extinguished and had disappeared. Third, the map itself represents a series of pathways and locations within Treblinka that testified to the utility not just of what its captives had brought with them on the trains but the very utility of their physical flesh for the industrial service of the Third Reich.

In some ways the 'Eternal Treblinka' of the animals differs entirely from that of the originary site of the human massacres. Its unmitigated animal massacres are of course on an entirely different quantitative scale to that of the human holocaust. The subjection of the Jews to sub-human status as bestial was about erasing them from the species of humanity, all the better to erase them physically. Animals had never entered the human taxonomic index. The disentanglement of humans from their animality had begun in the Palaeolithic era and had largely been secured by the advent of classical civilisation even if human society remained obsessed by its animal origins and its relationship to them (Hudson 2016a). It differs further from that originary Treblinka also in the fact of recognition. Holocaust denialism, refuted by the physical traces that Grossman himself witnessed and represented in his map, is untenable. Animal speciescide is still utterly denied. There are no maps of its location and indeed there could not be – it was a global phenomenon taking place in a multitude of locations. If any map had been created of this species-destruction it was designed by the hunters to make their subjugation of species and industry and inventiveness all the more effective. There were few logbooks that documented the scale of destruction and neither were there many witnesses who were not hostile to the variety of species being killed and sometimes as a species extinguished entirely.

If the use of the goods and the bodies of the captives in the originary Treblinka had some utility this was as a by-product of their destruction. The Jews and the Roma were killed because of their very genetic being as a perceived inferior species. The industrial utility of their remnants was simply a welcome contribution. The destruction of animal species was profoundly different. It rested on their very bestiality and their sub-human taxonomic status but their destruction was a consequence not of the hatred that humans bore towards that bestiality but almost purely because of utility. Utility was the central 'motive-power' of their destruction, partly sometimes to protect humans from wolves and lambs from birds of prey, but fundamentally because in their very corporeality the animals had something that humans, or some humans, needed.

The material, organic entities of animals had 'flesh' and human dominion over nature, justified by biblical and scientific tract, justified that that flesh was there for the pleasure of the human beings. Further, the extension of capitalism meant the extension of industrial slaughter upon the animal world and the use of vast networks of machinery to aid that subjugation. Far from subsistence hunting, this extension of industrialised slaughter meant the accumulation of the remnants and the disjecta membra of the captured species and thereby the accumulation of capital which was further invested in the ongoing industrialisation of nature and natural forms. At the same time, just like the appendages within the industrial machine, sentient human labour became antagonistic elements in that vast machinic system of subjugation – from the Atlantic whaling ships to the grand industrialised slaughterhouses of Chicago. Global transportation systems were developed to transfer flesh and the machines of industrial capitalism in the metropolis were fed by fuel from the captive species and illuminated by the same. In a way 'enlightenment' of industry rested upon the classification systems and taxonomies developed by Enlightenment science itself – as we can see in the work of William Scoresby the younger.

The human invasion

> Man has been a hunter for thousands of years. In the course of countless chases he learned to reconstruct the shapes and movements of his invisible prey from tracks on the ground, broken branches, excrement, tufts of hair, entangled feathers, stagnating odors. He learned to sniff out, record, interpret, and classify such infinitesimal traces as trails of spittle. He learned how to execute complex mental operations with lightning speed, in the depth of a forest or in a prairie with its hidden dangers.
>
> (Ginzburg 1990: 102)

Carlo Ginzburg has explored the notion of hunting as the search for *traces* (and as an analogy for historical practice). The co-habitation between humans and animals was often one of contestation in which each species sought the destruction of the other for survival and subsistence. Anthropologically this was often perceived in representations of brotherhood, respect and spirituality in which the spirit of the animal, as it was consumed by the human, has its spirit subsumed into the sentience of its eater. The extension of the technosphere of humanity and its hunting prostheses would develop into a vast system of species-destruction and accumulation with the ideologies of classification and zoology used to make the hunt more successful.

In fact this was a dual operation in which scientific classification and capital accumulation of species were co-produced for mutual benefit. The meagre hunting tools of stone knife and microlith would subsequently be replaced over multiple generations by the metropolitan machines consuming whale oil collected by highly weaponised whaling ships with a multiplicity of techniques and material tools designed to kill and carve the corporeal remnants of the Cetacean. The animal rights activist Ronnie Lee, often cited as an inspiration for the development of critical animal studies has pejoratively condemned the emergence of this machinic subjugation of species:

> It would do well for us to speak of human imperialism. Not content with just having its fair share of the planet, the human species has everywhere invaded and despoiled territories which rightfully belong to other creatures. Perhaps the worst words ever spoken (if, indeed, they were) were "Go forth and multiply". A call for a human occupation of the world similar to that of the Nazis for "Lebensraum". Thus the end of vivisection labs, of factory farms, will never be enough because it still leaves behind the injustice and oppression of the original "enemy" occupation. True animal liberation will not come merely through the destruction of the Dachaus and Buchenwalds that the occupiers have built for their victims, but demands nothing less than the driving back of the human species to pre-invasion boundaries.
>
> (Lee 1990: 1)

The destruction, in other words, of the techniques, locations and tools of the animal holocaust is not enough. The whole apparatus of human technocratic civilisation has to be eliminated and its human population radically reduced in order for species-restoration and a proper co-habitation of the earth to emerge again. The 'human movement' or 'human imperialism' as Lee calls it is based upon carnism and the devouring of populations 'other' to that of the human.

David Nibert condemned this 'predacious nature' of human socialities in his preface to a key work on critical animal studies (Nocella et al. 2014: x). For Nibert:

> Using horses as instruments of war and the salted flesh of other animals as rations, aggressive European elites dispatched their minions to ravage the Earth in search of wealth – much of which was in the form of the skins, hair, body fat, and tusks of other animals…The destructive invasions relied on state power, and the carnage was rationalized by the use of racism, sexism, speciesism, and other reprehensible ideologies. The resulting ill-gotten wealth allowed the rise of capitalism, a system birthed and continually nourished by the bloody, entangled oppression of the great mass of humans and other animals.
>
> (Nocella et al. 2014: x)

The carnistic (literally, 'flesh-devouring') ideologies of species domination by humans was not just about human 'nature' per se as if this was somehow part of what the human essence was. The domination was part of vast epochal and social forms which had emerged to exploit nature and profit from it in terms of material resources and capital. The accumulative power of mercantilism founds its richness in the accumulation of the body parts of animals. This was not destruction as a punishment for bestiality but a rational, accumulative mechanism justified by certain types of ideological emblems and motifs. Its 'motive-power' was capital and not emotion, although it could only succeed if the scale of nature could reduce the other species to dominion status and use-value. The development of the 'fixed-capital' of the machines was used to accumulate the 'mobile-capital', if we might call it that, of the hunted species. Much like the organic/inorganic hybrids of the metropolitan machine the hunted species was a form of capital. If the machine was indeed fixed-capital, the mobile pursued species was as much a form of potential capital as coal, or sandstone which even if organic in origin was now petrified. The living species however was not petrified and fossilised and actually had one further aspect to it that was attractive to capital expansion and accumulation – it could reproduce itself organically and potentially offer, and did offer, new generations to the hunter's spear.

The industrial slaughterhouse was, like all social machines of capitalism, a hybrid of the organic and the inorganic. As David Nibert says again – 'In Industrial slaughterhouses, which are perhaps the most nightmarish manifestation of the capitalist system, one commodity – the labour power of devalued people – is used to kill and convert other animals into other commodities that are foisted on the public as food', (2014: 14). The destruction of the animals benefited only one part of humanity. The segmented and fractured humanity of the social class system relegated other human beings as labour to an almost bestial status (and often entirely bestialised its labour as in the Nazi and Soviet death camps where the human 'beasts' laboured for the industrial enterprise of the whole despotic social and military system).

In their critical analysis of anthropocentric 'carnism' and the extension of the human into nature, Weitzenfeld and Joy (2014) have argued that it is a 'distorted humanism':

> anthropocentrism, which has narcissistically privileged humans as the center of all significance, is not an innate disposition but a historical outcome of a distorted humanism in which human freedom is founded upon the unfreedom of human and animal others. Nonetheless, the humanistic aspiration to ground knowledge in reason through modern science has undermined the tenability of a human-animal binary and hierarchy, thus resulting in a critique of speciesist beliefs and institutions.
>
> (Weitzenfeld and Joy 2014: 3)

The ideologies of this distorted humanism, based around conceptions of edibility of the animal 'others' is, they argue, catastrophic. The notion of centralisation that we have seen in terms of capital, machines and population finds its parallel in humans who accrue to themselves as the repository of all value and meaning the products (and edibles) of the natural forms (Weitzenfeld and Joy 2014: 3–4). If the 'distorted humanism' of carnism is based upon a sense that only the humans are 'self-forming' and 'world-forming' and therefore in a sense can transcend their 'natural forms' then the opposite holds – the essentially static animal is bound by natural law and therefore does not have the same moral status as a human being. The animal exists as a 'nature' which can be inscribed upon by these humans in self-transformation – they are simply instrumental entities subject to the will of the sentient human masters. The 'human-animal binary' confirms the supremacy of the human being and its orders, scales and continuums of nature reaffirms hierarchies which 'continue to reproduce a bloody margin of subhumans' (Weitzenfeld and Joy 2014: 6).

In fact the very status of the human 'being' itself is problematic. In their reading of Agamben, Weitzenfeld and Joy describe the human as 'not so much a value-neutral biological fact as a violent political fiction' (2014: 8). These fictions, or concatenations of violent discourses, destroy humans just as much as they destroy other species. The 'auto-vivisection' of the human psyche separates human from themselves and from all other beings in the world (Bell 2011: 166). The metabolic rift then is not just between humans and nature, humans and animals, humans and machines but present within the interior of the human itself.

The Cetacean holocaust

Heathcote Williams, in a work which perhaps marked a cultural sea-change in animal studies, has argued for an understanding of whales as families, tribes and nations. What is distinctive about whales is their very archaicism with their 'Cetacean music' expressing the memory of millennia (1988: 17). Williams argues aesthetically for the sentience of the whale tribe and the 'Fifty-million-year-old sagas of continuous whale mind' (1988: 19). He, too, makes analogies with the holocaust and remarks that human civilisation's 'foundations were built upon dead blubber' (1988: 95). The final section of the book 'On the Nature of Whales' is a collection of folklore and science of the Cetaceans which in a sense becomes the human gesture of archiving of a species which they have diminished (1988: 111–182).

Young's vertebrate taxonomy locates the 'plasticity' of the animal form of whales, arguing for an understanding of their 'fish-form' as a reversion. He notes that 'The whales are a successful set of populations of aquatic, carnivorous mammals, which diverged from the eutherian stock, possibly from the creodont branch, at a very remote date' (1962: 666). For Young, like Williams, the whales are defined by their sentience and their sociality:

> The behaviour of whales is undoubtedly elaborate, involving social life, communication by sound, and probably much learning. There is cooperation between individuals in helping to keep a wounded companion or a new-born at the surface. Play is common and rhythmical 'dancing' has been observed and also homosexual behaviour, in captivity.
>
> (Young 1962: 672–673)

These are not anthropocentric impositions on the whale but the elucidation of characteristics that both whales and humans share. Of course the former did not share the very characteristic that would bring the whales to the verge of extinction – that of accumulation, of carnism.

Gaskin's history of the seaport of Whitby, on the coast of North-East Yorkshire, describes the whale hunters and the captives that they brought back – already cut and vivisected on the decks of the flensing ships.

> The moment of triumph came when, with the barbed harpoon buried in its quivering side, the mighty animal was towed to the ship by the merry procession of boats. After followed the slinging of the carcase alongside; the flensing, or cutting-up into slips, with large knives; the hoisting of these junks of blubber on deck, whilst the cheery refrain went over the waters…
>
> (Gaskin 1909: 259)

Like the slave ships plying their trade in other parts of the Atlantic world (Hudson 2016b), the whaling ships were machines designed to secure captives of non-humans. Unlike the slave ships, the North Atlantic whalers had no intention in keeping the captive cargo alive. The value of the product lay not in the congealed labour of the captives which could be employed in the new agricultural and industrial systems of proto-capitalism and capitalism, but literally in the congealed flesh of the material body of the whale.

In the same way as the slave ship acts as a 'chronotope' for the modernity of the 'Black Atlantic' (Gilroy 1993) so the whaling ship acts as an emblem and motif of the 'Cetacean Atlantic'. This is not to conflate those processes, which were profoundly different, nor to downplay the Jewish or the African holocaust. The opposite in fact. Understanding the speciescide of other beings can help illuminate the capacity of humans for violence against their own species particularly the process by which one living entity is relegated from the scales and hierarchies of nature. The clues to the meaning of all of our holocausts lie in the recognition that each of them was unique and had different and complex motive-forces. But enmeshed and entangled in each of them was a set of industrial, machinic structures that were designed to subjugate, transform and ultimately destroy others. The map of Treblinka can guide us through the middle passage of the slaves and across the polar night to the fabled North West Passage of the whalers. Ultimately, it offers solidarity to the victims of those huge, capitalised engines of destruction.

Scoresby and the Greenland trade

The emergence of the whaling ports of England, Scotland and the Scandinavian world were propelled by the proto-capitalist development of accumulation from the late medieval period onwards. The oil of the destroyed whales drove the machines in the early factories. One commentator on the whaling industry noted that 'Every new piece of machinery needed lubricants, every household needed lights…' (Barrow 2001: 65). Whale oil was used in the lighting of mines, workshops and factories, as an 'illuminant'. It was also used in soap and paint (Barrow 2001: 65). The volume and density of the tools and machines used in the trade demonstrated its value and ships and their masters would compete for how many whales they could land on each trip. Foremost amongst the whalers were the Scoresbys, father and son, whose work (largely documented by the naturalist and explorer William Scoresby junior) involved the ethnography and observation of whales alongside their killing. This history of whaling is expressed in the poetry of R.S. Thomas who writes of a great stomach where bones, wrecks and continents are contained but also displayed (see Causley 1978: 23).

William Scoresby senior, was particularly famous for the theatrical tableaux that he would perform over the corpses of whales with his sailors as willing and cheering audiences and participants. It is worth examining in some depth the son's vision of the father. Scoresby senior, was both whaler, farmer and inventor of the top-gallant 'crow's nest'. His voyages to the Polar north were legendary during the whaling period, not just for his innovations, but for the scale of the accumulation that he set his industry towards. As his son says:

> To the requirements, however, in every species of knowledge and duty connected with the Arctic navigation and the capture of the huge cetacean of the north, he gave himself with such tact and perseverance, that, on his sixth voyage, we find him to have risen over the heads of all his original associates, and occupying the position of second officer, the *specksioneer* of the ship.
>
> (Scoresby 1985: 37)

The 'specksioneer' was essentially the second in command of the ship and in charge of one central process: the death and then the cutting up of the captive whale. Scoresby junior revels in the quantitative vortex of the whale ship's accumulative power. On his father's second voyage as captain, he notes that he took eighteen whales yielding 112 'tuns' of oil (1985: 55). The metrics of the trade were truly advanced and it took place essentially on an industrial scale – each part of the whale had its utility for fuel or for food and its disjecta membra, that which could not be used, was toppled into the sea.

Yet the story of another of his father's captives, and one which survived for a period, illuminates the kinds of discourses which made the subjection of animals not just valued but necessary. This lies in the retelling of the story of Scoresby's polar bear (Scoresby 1985: 78–85). On one of his journeys, Scoresby senior captured a polar bear and chained it as a captive upon the decks of his ship. Allowed by God to torment the bear as a symbol of the human dominion over nature, Scoresby initiated a programme over the bear to induce it to dread and fear humans. Through the violence inflicted upon the enslaved bear and then through calming and stroking it, Scoresby gradually broke the will of the bear and tamed it (1985: 85). As his son proudly made clear: 'The thorough subjection, indeed, of this naturally ferocious creature was soon effected, I believe, of two or three days, – and from that time forward my Father's command over it was uniform and supreme', (1985: 80).

After landing the bear at Whitby and chaining it up in his oil yards by the side of the River Esk from whence it escaped, only for it to be recaptured by its master who calmed and soothed it as a pet, Scoresby tired of the bear and had it transported to the Tower of London and its menagerie of 'wild beasts' (1985: 83). A year later, on a tour around the Tower he visited the Polar Bear in its cage who recognised its old master and submitted itself to his touch as if he were a passing king (1985: 84). This was, for both father and son, a subjection of beasts which was ordained by God. As Scoresby notes in his lesson about his father and the bear: 'It was His design and command, in respect to the inferior creatures, that this should be so...But the appointment, which was simple and natural when all was innocency, was afterwards renewed, we find, under a *new influence*, that of *fear*, specially induced on the general constitution of the animal creation' (1985: 84). The initiation of fear as a form of control would anticipate the totalitarian politics of the twentieth century and in the context of the eighteenth century was parallel to the development of the 'Thanatocratic' British state (Hay et al. 1976, Linebaugh 1991, Thompson 1976). This was a state ruled and disciplined by the fear of death – a disciplinary regime which had its first extensive manifestation in the persecution of the vagabonds during the reign of Henry VIII (Marx 1964). One important aspect of the Thanatocracy, whether on board ship or in the squares of the cities, was the act of display in order to induce fear and discipline. The struggle against the bears would continue with almost a war conducted against them on the *Baffin* voyage of 1820 (Scoresby 1823: 109–114).

Scoresby, writing of his father's display of his virtuosic 'flensing' skills, revels in the details of the dismantling of the whale and the tools and machines that were used to do this. Scoresby junior, later documents the actual capture of a whale at his father's hands and weaponry:

> Lanced, now, on both sides at once, with these formidable instruments of destruction, the dangerous energies of this vast animal soon become overpowered, and it now yields itself passively to its inevitable fate. One effort alone remains. The instinctive impulse or spasm of expiring vitality, – like the brilliant gleam or coruscation of the expiring taper, is expended in a series of tremendous death-throes.
>
> (Scoresby 1985: 148)

Once the great whale has been conquered and extracted from the sea and tackled to the side of the ships, the dead animal is transformed into a corpse to be vivisected and transformed into capital. For Scoresby, examining the operations of his father, the death of the animal is only the beginning of its transformation:

> The tardy formality with which the "flensing" of the whale was accomplished, irrespective of the particular magnitude of the animal to be despoiled of its blubber and whalebone, was frequently a source of great annoyance to my Father. The number of cuts, with the placing of straps, and attachment of tackles, had become – like the skeleton forms issued by public offices – an established system; and, cumbrous as it was, with respect to fish of smaller growth, it was made generally applicable to all. The effect of this was, that whilst the largest sized fish would be flensed in about four hours, the taking in of one of the fourth or sixth part its size would occupy nearly half as much time. An hour and a half at least, but more frequently two hours, at that period of the fishery, would be expedited upon one of the ordinary small-sized whales. The poor little carcass, indeed, was encumbered by the number of the harponeers (to whose province belongs the fixing of the tackles and the cutting away of the blubber) congregated upon, or about it, whilst flensing.
>
> (Scoresby 1985: 108–109)

In order to demonstrate the industrial efficiency of his work, the elder Scoresby prepares the display of his operation around the material corpse of the captive whale. Each sailor becomes an appendage in the mechanical dismantling of the structure of the whale encumbered as it was by ropes and tackles as its skin and blubber is first cut away. His son continues the narrative:

> Opportunities for the experiment being at this time abundantly afforded, he forthwith prepared himself for this trial of skill. The available hands – that is, excluding cook, steward, surgeon, etc. – were usually about forty-five or forty-six in number. Out of these he took, not a picked set, but only two boats' crews, with their supernumeraries, according to their existing classification, comprising about sixteen men. These he appointed to their several stations about the deck; eight to the capstan, four perhaps to the "crab" or "winch", and the rest to manage the "tackle-falls", to cut the blubber up and heave it into the "flensgut", or receptacle for it below.
>
> (Scoresby 1985: 109–110)

Scoresby then prepares his instruments – blubber-spade and blubber-knife all ground, sharpened and polished and noting the time the captain begins his challenge and demonstration (1985: 110). The head of the whale is taken on to the deck, the fins are 'unsocketed' and the 'bold experiment' begins (1985: 111–114). The 'master-hand' dismantles the whale, upon the decks of the whale-ship *Dundee* in record time: 'As fast as the men on deck could heave-up the blubber, the blubber was freed to their hands. Every change of tackle, or place of working, was so managed as to leave no interruption in the labours of his men…An instinctive spirit was infused into every department; for no section of the men liked to be behind, so as to be in the humiliating position of hinderers of the others', (1985: 114). Each microscopic aspect of the ship is examined, the use of tackle, the turn of a blade, the heave of a sailor's arm. Each is part of the complex armature of the ship all dedicated to the destruction of the whale and to the accumulation of its elements which will rapidly be turned into capital from the congealed potential capital of the blubber and the oil. Each element of the whale is met with an elemental part of the ship's machinery and the organic human hand of the master wielding his patriarchal power over the body. The cuts of the mechanical tools are directed towards the membrane and each part is geometrically separated from the other according to a model and template that had been worked out from a generation of 'flensing'. For Scoresby:

> His sharp and finely-polished *spade* with which he chiefly worked, seemed to meet with no resistance from the animal textures against which its edge was directed. Instead of cutting downward through the blubber a spade-breadth at a time, as most usually was done, he would run the instrument in a direction obliquely horizontal, so as to separate the slip

> heaving up from the general envelopment of the blubber, for a yard or two in extent, at a single stroke or thrust of the instrument. The slight attachment of the blubber to the muscles of the carcass could then be usually torn away (with a touch underneath with the spade when incidentally needful to be cut) by the force with which the slip was being raised by the mechanical engines in motion on the deck.
>
> (Scoresby 1985: 114–115)

Any residual resistance was of course not from the sentience of the whale, now dead, but from the resistance to the tools in its material being. In all other ways it had been subjected to the dominion of its human master and the machine of the ship.

The performance demonstrated here is perhaps worth examining in explicitly theatrical terms. Scoresby has accumulated his audience as he has gathered his sailors into the ship from the northern ports, including the north of Scotland, and from his own home fields. They are gathered together upon the stage of the ship in order to witness the contest between their captain and the dead body of the whale. Scoresby speaks to the sailors, addresses them, encourages them to admire him. He wields his 'props' around his body and upon the whale. There is no fourth wall to break out of. There is no pretence of separation between the realist, naturalistic performance because it is indeed realist and part of God's ordering of nature (See Goode 2015 for the dissolution of the wall and theatrical space and place). Like the anatomical performances over the human body performed in the medical schools of the same period, Scoresby's 'act' is one of disarticulation, clinical knowledge and also of 'utility'. Scoresby's 'dominion' is one that is both ideological and physical – the part of nature that he oversees and controls includes both the body of the whale and the ocean from which it is extracted. The performance is at once the display of human mastery, of God's natural order and of Enlightenment scientific exploration and explanation. But it is also a performance of ideological and physical power over the labouring humans on his decks and an affirmation of the total rule and mastery of the captain.

Each invention of Scoresby, whether for navigation, observation, and flensing increased the ease of submission of the whales and the efficiency of the ships. His most radical invention was the development by May 1807 of the 'crow's nest' perhaps suggested by being lost in the moorland snow in his youth (1985: 137). Labour is managed in minute ways. The ship was an encampment of the labouring proletariat, its labour divided both by task and by hierarchy. But Scoresby's labourers were also part not just of the nautical proletariat but the agricultural proletariat of his land. The seasonal whale voyages were largely combined with work on the fields when ashore and this was just as subject to Scoresby's industry and invention as the ship's technical inventiveness. The ship itself, built of wood and iron, driven by wind, manned by human labour was half organic and half inorganic. Its machinery subsumed labour, separated its labour both from others and from the interior life of the proletarian.

If the 'wooden world' of the ships of the Atlantic slave and whale trade reflected the social composition of wider English society, as N.A.M. Rodger correctly argues in his study of the English navy, they were also in contra-distinction to Rodger's thesis 'Floating concentration camps' (Rodger 1988: 344–346). They were camps for the indentured labour of the sailors but even more so were part of the 'Treblinka' of the Cetacean holocaust. They concentrated, like the industrial metropolis, labour, capital and machine on their decks. They also concentrated the captured whales, once mobile and free, upon their decks – displaying them as a form of power and accumulating their parts for capital.

Peter Linebaugh, in his analysis of the emergence of the Atlantic working class, has argued of the Atlantic ship that 'All the contradictions of social antagonism were concentrated within its timbers' (1982: 109). The vast 'circulations of experience' of the Atlantic nautical proletariat would itself eventually come to threaten the capitalism that had brought into being. The ships were perhaps the highest expression of the machinic accumulation of capital and the possibilities that the ocean opened up for both capital and wage-labour. For Linebaugh:

> The large capital outlay, the division of labour, the regimentation and repetition, the close supervision, the working in groups, and the removal from home are the characteristics sailoring had with the factory. The ships' holds carried the congealed labour of the plantations, the manufactories, and the workshops to their new destinations. The rolling and pitching deck was the uncertain stage upon which the international seafaring proletariat told its stories to one another and sought for itself its historic role.
>
> (Linebaugh 1982: 108)

The 'Thanatocratic' weaponised ship was indeed the product of the social formation of capital. The 'flight of the serfs' from the fields and into the factories and ships that Marx noted in early capitalism would create new forms of proletarian experience just as the empire of capital began to subject and consume the remnants of other species in order that its own mercantile forms of circulation were concentrated, sustained and intensified.

Tom and Cordelia Stamp, in their various studies of Scoresby, have described the subjection of the crew to the master on his ships and the conditioning and division of human labour. They note: 'So, at any given time aboard a whaler was found this body of men in various transitional stages of development and experience…' (1983: 13). The development of human labour rested on the hierarchical relationship to the command and one's position within the machine of the ship. There were both individual and social mutinies upon the ships and a variety of different ways of disrupting its mechanism by the human appendages. The disciplinary regime of the ship was also disrupted by attempts to seize it (1983: 22–24). In one case the mutineer John Wright, suspended the entire flensing operations of the ship and threatened to have 'blood for supper' except he would not deign to taste the captain's blood (1983: 23). Scoresby himself noted the array of human labour on his ships: 'The crew of a whale ship are separated into divisions' with labour divided and allotted to different boats (1969b: 234). The fortifications that were built on the ships protected it from both the whales and against ice (Scoresby 1823: 307). Its array of weaponry and tools was as useful to fortify the ships against mutiny as to hunt the whales. Scoresby himself listed a full directory of apparatus and machinery on the ships (1969b: 509–511).

Scientific researches into snow, ice, wind, navigation, rocks, plants, species were also a major part of Scoresby's expeditions again reaffirming the relationship between naturalistic observation techniques and species destruction in the voyages of ships such as the *Baffin* (Scoresby 1823). Scoresby's analysis of the mirages of human cities that he saw upon the littoral of the Arctic ice sheets (1823) would in turn become a central part of H.P. Lovecraft's (an avid reader of Scoresby as was Melville) cosmology of the icy darkness of Antarctica (2007). Scoresby's zoological overview of the Arctic regions was not intended by him to be a 'systematic Fauna Arctica' (1969a: 446) but a guide whose ultimate purpose was to epistemologically and practically navigate his own ships to hunt, capture and kill large parts of the Cetacean species.

Noting the inconsistency of textual descriptions and visual representations of the Cetacea (1969a: 447) Scoresby observes the whale in a highly specific manner: as almost an act of complete 'deconstruction'. He observes the whale in its ocean not as an independent, living entity in its world, but as something which can be looked through as if its membrane was invisible and its structure, parts and oil could be perceived beneath and ready to be utilised. For Scoresby the Greenland Whale is representative of the species as a whole:

> This valuable and interesting animal, generally called *The Whale* by way of eminence, is the object of our most important commerce to the Polar Seas, – is productive of more oil than any other of the Cetacea, and, being less active, slower in its motion, and more timid than any other of the kind, of similar or nearly similar magnitude, is more easily captured.
>
> (Scoresby 1969a: 449–450)

Scoresby, examining the whale, begins to unpeel its various membranes and its blubber: 'The whale has no external ear; nor can any orifice for the admission of sound be discovered until the skin is removed' (1969a: 456). He goes on to examine not just the science of the blubber but methodologies of extraction of oil from it and from the porous bones of the whale (1969a: 460–463).

Later in life, and in comfortable retirement, there are accounts that Scoresby's 'extreme sensitivity' led to his repugnance at the killing of whales (Stamp and Stamp 1975: 226). Indeed, he expresses this in what we might call his anatomical guide to the whale's destruction:

> There is something extremely painful in the destruction of a whale, when thus evincing a degree of affectionate regard for its offspring, that would do honour to the superior intelligence of human beings; yet the object of the adventure, the value of the prize, the joy of the capture, cannot be sacrificed to feelings of compassion
>
> (Scoresby 1969a: 472)

His own palate had a taste for whale flesh although he recognised that it did not entirely suit the stomachs of Europeans. The body of the dead whale is at once a source of meat and a scientific laboratory in combination – '…the flesh of young whales, I know from experiment, is by no means indifferent food' (1969a: 475–476). He notes that the whales had few enemies and adversaries and was so timid that even a bird alighting upon its back could agitate and terrify it (1969a: 474). This timidity suited its central adversaries – the human beings of the North Atlantic world. The *Eclectic Review* in their review of Franklin's natural history of the Arctic and the whale deplored the 'Munchausen-like engraving of a whale playing at battledore and shuttlecock with a boat and its crew' (Stamp and Stamp 1975: 80). This is a particularly striking image of the human-animal warfare of the Atlantic passages. It displays the whole machinery of animal repression and the resistance that the great whale offers to human mechanics, logics and planning.

Scoresby the naturalist was enmeshed in the scientific Enlightenment of England and one of his closest associates was Joseph Banks (Stamp and Stamp 1975: 32–33). By his old age he was reckoned to be the foremost commentator on the North Atlantic world and as such was enlisted by the bereft Lady Jane Franklin to help map the territories in which the Franklin expedition had disappeared (Stamp and Stamp 1975: 209–210). Yet it is the political economy of whaling that demonstrates the horror of those contact zones where the species met. As Scoresby noted:

> We are led to reflect on the economy manifest in respect to the hugest of the animal creation, whether on earth or in the ocean, whereby all become subject to man, either for living energy or the produce of their dead carcasses.
>
> (Stamp and Stamp 1975: 71)

In his chronological history of the whale trade Scoresby arithmetically elucidates the quantitative mass of whale flesh that he and his ships were accumulating in terms of tonnage (1969b: 119–130). Every part of the whale was used, even the 'finks' and 'fritters' of the skin that were used to boil other parts of the whale (1969b: 176) and the industrial process of the flensing (1969b: 298–301). He also, in attempting to map the routes of the fleeing whales, considered them as living in tribal structures, each tribe living in different regions of the ocean and having differing migratory routes (1969b: 209–211). Recognising the 'social' nature of the whale meant the recognition of a social form adequate to subdue those tribes – the assemblage and the machinery of the ships.

Perhaps the most fascinating part of Scoresby's natural history and 'anatomy' of the whale were the specific uses to which the parts of the whale were put. This would bring the Scoresby family undreamt of riches once their cargoes had been flensed and packaged ready for transportation out of the seaport (1969b: 435–437). Whalebone was much sought after as it offered a structural 'skeleton' for items such as hats, umbrellas and sometimes field gates. But its most important use was in the stays of the whalebone corsets (1969b: 435). The development and enlargement of the hooped dress was partly to displace a reliance on the weight of multiple petticoats from the seventeenth century onwards but as Aileen Ribeiro has argued in her analysis of the 'extremes' of hoops, they were becoming larger and more structural (2002: 42–43). As Ribeiro notes: 'The first hoops were roughly in the shape of a birdcage or rounded dome...They were made of linen or stiffened canvas with rows of whalebone (wood was a cheaper alternative); the English hoops had eight tiers, the French usually five', (2002: 43). One commentator apparently noted that at least one dress was based upon St Paul's Cathedral. What is remarkable is that as the circumference of dresses increased, with multiple structural hoops around the womens' bodies, they appeared to replicate the structure of the whales themselves. The 'stay-makers', most often men, would visit the 'ladies' in their own homes and bring the whalebone with them to design the structure of the skeleton around the body of the specific woman (Ribeiro 2002: 73) and working-women often used either leather or canvas 'boned' with reeds to replicate the whalebone they could not afford (Ribeiro 2002: 85). Ribeiro notes that 'By the middle of the 1720s, the hoop had assumed an oval shape, sometimes more than eleven feet in circumference, and by 1730 the front of the skirt was flattened by a system of cords inside, which made the hoop more manoeuvrable and more decent when the occupant sat down. As it flattened in the front, so it dramatically increased in width at the sides during the 1730s and 1740s, causing problems during movement (women had to go sideways through doors) and in a confined space', (Ribeiro 2002: 43).

The body of the woman encased as it was in the birdcage, the cathedral, the ship (all metaphors that were used for the dress) had, in a literal sense, to navigate her own routes around confined spaces with a network of pulleys aiding her passage. Yet the whalebone structure also replicated the whales themselves: silk replacing the membrane of skin and blubber around the skeletal frame. Their carcases ransacked and stripped, the material entity of bone finds new architectural expression imprisoning the body of the woman and hiding her physical, material frame from passing observers.

Melville's modernity

The 'indomitable excess of the dialectic that both Marx and Melville see incarnated in the living flesh of labor' (Casarino 2002: xx–xxi) is of course also inscribed literally into the flesh of the species which capital has subdued. Nowhere is this obsession to accumulate or destroy more present that in the incarnation of the hunt represented by Captain Ahab in Melville's *Moby-Dick* first published in 1851 (1992). There is a ubiquity of commentary on Melville in literary theory including work on his natural taxonomies, his racialisation, his modernistic literary devices and tropes and his unique geographical knowledge. Here we focus on the relationship between the ship as machine and the Cetacean species in order to understand how Melville represented industrial modernity through the microcosm of the whaling ship *Pequod*. Melville sees the archive of domination in both the history of the whaling ship and the body of the whale itself:

> He swam the seas before the continents broke water; he once swam over the site of the Tuileries, and Windsor Castle, and the Kremlin. In Noah's flood he despised Noah's Ark; and if ever the world is to be again flooded, like the Netherlands, to kill off its rats, then the eternal whale will still survive, and rearing upon the topmost crest of the equatorial flood, spout his frothed defiance to the skies.
>
> (Melville 1992: 504)

Melville's aesthetic representation of the 'wooden world' of the *Pequod* is itself a representation of the *Pequod's* aesthetics – specifically its aesthetics of slaughter. His initial description of the ship, narrated by Ishmael, makes observations upon the ship as an archaic and antique entity:

> She was a ship of the old school, rather small if anything; with an old fashioned claw-footed look about her. Long seasoned and weather-stained in the typhoons and calms of all four oceans, her old hull's complexion was darkened like a French grenadier's, who has alike fought in Egypt and Siberia. Her venerable bows looked bearded. Her masts… stood stiffly up like the spines of the three old kings of Cologne. Her ancient decks were worn and wrinkled, like the pilgrim-worshipped flag-stone in Canterbury Cathedral where Becket bled. But to all these her old antiquities, were added new and marvellous features, pertaining to the wild business that for more than half a century she had followed.
>
> (Melville 1992: 77)

The archaic 'essence' of the machine of the ship is analogical to ancient human beings and old flag-stone – the organic wooden plans of the ship literally weathered by storm and tide. To the 'old antiquities' of the core of the ship's machine were added new organic features both of ornamentation and decoration as well as entangled with the machinery of the ship. The whole of the *Pequod* was literally bedecked with the skeletal remnants of the whales that the ship and its captain had destroyed. The industrial slaughterhouse aesthetic created a reliquary of bones strung around the ship:

> She was apparelled like any barbaric Ethiopian emperor, his neck heavy with pendants of polished ivory. She was a thing of trophies. A cannibal of a craft, tricking herself forth in the chased bones of her enemies. All round, her unpanelled, open bulwarks were garnished like one continuous jaw, with the long sharp teeth of the sperm whale, inserted there for pins, to fasten her old hempen thews and tendons too. Those thews ran not through base blocks of land wood, but deftly travelled over sheaves of sea-ivory. Scorning a turnstile wheel at her reverend helm, she sported there a tiller; and that tiller was in one mass, curiously carved from the long narrow lower jaw of her hereditary foe.
>
> (Melville 1992: 78)

The machinery of the ship, pinning mechanisms and steering mechanisms, is literally carved from the 'sea-ivory' of the whale corpses. Like the great jaws of whale-bone that often bedecked the rural farmsteads of old whale-sailors these were trophies of subjugation but also of 'experience'. Philip Hoare has noted that this history is not just about the whale or the ship but about the entirety of life and of human and animal existence: 'Man, whale, life, death: this was the story Melville had to tell' (Hoare 2009: 173).

The reliquaries were literally an archive of the history of the ship, its captives and a material representation of the social memory of the nautical proletariat. It signified both the supremacy of the machine and represented in material form the apotheosis of the organic version of the machine – built from dead forests and exterminated whales. It also expressed the material template of the dominion of fear – a Thanatocratic warning to any species that would cross its wake. The antagonistic social struggles on board the ship (Bryant 1998: 68) also reflected the ideologies and struggles of the wider industrialising world. The monstrous ship's master, what John Bryant calls the 'Shakespeareanized Ahab', may have even been based on Shelley's Frankenstein (Bryant 1998: 67). The dissolution of the narrator Ishmael as a character, creates of him a new archivist of the Cetacean species. As John Bryant notes, 'Nevertheless, as his role as a character erodes, his life as a lyrical, poetic meditator upon whales and whaling transforms the novel once again, converting Ahab's drama into a vast essay on things cetological', (Bryant 1998: 67–68).

The material culture of the ship was also comprised of a 'multitude of things' including spare lines, extra harpoons and extra boats largely because of the threat of the material destruction of the trade and the catastrophes that the hunt might face (1992: 104–105). Melville makes the analogy between the butchery of military commanders and the butchery of the whale decks. The 'defilements' of the trade are indeed vile but 'what disordered slippery decks of a whaleship are comparable to the unspeakable carrion of those battlefields from which so many soldiers return to drink in all ladies' plaudits?' (1992: 118–119). The lamps of the whole globe, argues Melville, are shrines to the glory of whaling and the extension of geographical knowledge of the earth unthinkable without its navigations and ships (1992: 119–120). The whaler himself, or at least the captains and specksioneers of the ships, were the masters of species and of nature: 'I know a man that, in his lifetime, has taken three hundred and fifty whales. I account that man more honourable than that great captain of antiquity who boasted of taking as many walled towns', (1992: 122).

What is remarkable about Melville's account, through the narrative voice of Ishmael is the irony that the unbounded extinction of nature is at once a destruction based upon the extension of knowledge – a 'Cetology' or 'science of whales'. Melville notes the example of Scoresby whose zoological taxonomies of the Cetacean species made them all the more destructible (145–146). In fact a central part of the origin of the story of Moby-Dick was in Melville's reading of Scoresby (Bryant 1998: 65) as well as the documented history of the actual 'Mocha Dick' (see Rogers 1998). Even the geological and prehistoric observation of the petrified remnants of the Cetacean fossils illuminated new understandings of time and even proto-evolutionary theory, combined here with a form of biblical literalism:

> while in the earlier geological strata there are found the fossils of monsters now almost completely extinct; the subsequent relics discovered in what are called the Tertiary formations seem the connecting, or at any rate intercepted links, between the antechronical creatures, and those whose remote posterity are said to have entered the Ark; all the Fossil Whales hitherto discovered belong to the Tertiary period, which is the last preceding the superficial formations. And though none of them precisely answer to any known species of the present time, they are yet sufficiently akin to them in general respects, to justify their taking rank as Cetacean fossils.
>
> (Melville 1992: 497)

The metrics, speciation, dimensions and classifications of what Melville calls the 'leviathanic brotherhood' (1992: 148–151) were really an ethnography of a whole species at a particular moment. The observer/ethnographer and hunter are a single entity observing and 'capturing' the whale at the same time as killing it. The 'Cetacean science' is at one with the 'Cetacean holocaust'. The destruction of each whale was, like Scoresby's dismantling and autopsy of the carcases, a moment of ritual for the ship – 'The long, barbed steel goblets were lifted; and to cries and maledictions against the white whale, the spirits were simultaneously quaffed down with a hiss', (1992: 181). The crew 'of mortal renegades, and castaways, and cannibals' relished their role in the destruction of the animals (1992: 203).

Part of the horror that the sailors felt toward the whale lay in its deathly 'whiteness'. Like the polar bear that Scoresby brought into captivity, or the 'Xok' of the mid-Atlantic (see Rediker 2008), the sailors were appalled by its very colour: 'Witness the white bear of the poles, and the white shark of the tropics; what but their smooth, flaky whiteness makes them the transcendent horrors they are?' (1992: 205). The social and sailing imaginaries of 'charts' (1992: 215) and the navigation of the earth's circumference created new forms of knowledge which only enhanced the horror and the awesome nature of the whale. The ships were bedecked with the bleached, white trophies of the destroyed creatures. Melville becomes not just the archivist of the wooden worlds of the ships but of the disappearing Cetacean species. To archive and destroy at the same moment on the longitudes and latitudes of the north Atlantic. Melville's representation was a representation of actual experience of the Leviathan:

> So assured, indeed, is the fact concerning the periodicalness of the sperm whale's resorting to given waters, that many hunters believe that, could he be closely observed and studied throughout the world; were the logs for one voyage of the entire whale fleet carefully collated, then the migrations of the sperm whale would be found to correspond in invariability to those of the herring-shoals or the flights of swallows. On this hint, attempts have been made to construct elaborate charts of the sperm whale...Besides, when making passage from one feeding-ground to another, the sperm whales, guided by some infallible instinct – say, rather, secret intelligence from the Deity – mostly swim in *veins*, as they are called; continuing their way along a given ocean-line with such undeviating exactitude, that no ship ever sailed her course, by any chart, with one tithe of such marvellous precision.
>
> (Melville 1992: 217).

The task of the ship was to indeed follow in the *veins* of the whale passages in order to bring them into captivity. Understanding the migratory routes of the whales was a matter of observation and of course the replication of those migratory routes by the mobile factory of the ships was itself a new form of seasonal human migration or transhumance. In this case not in order to protect and keep the flock until their eventual slaughter, but to make it captive and kill it on first contact and encounter. The whalers were not shepherds but hunters. Melville himself remarks on the irony of eating one's food of whale meat under the light burning with whale oil (1992: 325). The ubiquitous use of the whale meat meant that following those migratory veins meant riches beyond just the accumulation of flesh.

The ritual capture and demolition of the whale was also entangled with oceanic superstitions about its spirit. Even as its outer membrane was pierced, its spirit continued: 'Desecrated as the body is, a vengeful ghost survives and hovers over it to scare' (1992: 336). But the immateriality of the whale's spirit is no barrier to the accumulation of its capital, its power and resistance has already been dispersed: 'Thus, while in life the great whale's body may have been a real terror to his foes, in his death his ghost becomes a powerless panic to a world' (1992: 337). Let us again look at the human and machine operations around and in the body of the whale:

> As in decapitating the whale, the operator's instrument is brought close to the spot where an entrance is subsequently forced into the spermaceti magazine; he has, therefore, to be uncommonly heedful, lest a careless, untimely stroke should invade the sanctuary and wastingly let out its invaluable contents. It is this decapitated end of the head, also, which is at last elevated out of the water, and retained in that position by the enormous cutting tackles, whose hempen combinations, on one side, make quite a wilderness of ropes in that quarter.
>
> (Melville 1992: 372)

The magnitude of the carcase meant dismemberment on an industrial scale and the ships were highly industrialised engines. Long periods at sea meant that dismantling the whale before its carcase rotted was necessary and also to make room for more whales, or whale parts, within its decks. The American ships had factory-structures which replicated those on land – particularly the 'Try-Works' – large brick kilns which were used to burn the carcases:

> Removing the fire-board from the front of the try-works, the bare masonry of that side is exposed, penetrated by the two iron mouths of the furnaces, directly underneath the pots. These mouths are fitted with heavy doors of iron. The intense heat of the fire is prevented from communicating itself to the deck, by means of a shallow reservoir extending under the entire inclosed surface of the works. By a tunnel inserted at the rear, this reservoir is kept replenished with water as fast as it evaporates. There are no external chimneys; they open direct from the rear wall… These fritters feed the flames. Like a plethoric burning martyr, or a self-consuming misanthrope, once ignited, the whale supplies his own fuel and burns by his own body.
>
> (Melville 1992: 462)

The ship, destructive as it was in its entirety, was only a machine of capture. It was the Try-Works which were emblematic of the whale's complete destruction and dissolution. The fuel from the whale's own body consumed its remnants and so 'the rushing Pequod, freighted with savages, and laden with fire, and burning a corpse, and plunging into that blackness of darkness' (1992: 463). The 'departed species' of the whale (1992: 498) is transported south across the latitudes of the Atlantic night to fuel further conflagrations in the homes and factories and machines of the 'human' world.

4 The human encampments

In the fourth chapter, we examine the extension of the 'camps' of humanity; its encampments, domains and cities. We look at the strata and sedimentation of the human 'occupation' of nature and territory. The human 'camp' was often an enclave which was mobile and in movement across territory either for pastoral or military purposes. The development of permanent occupations transformed human cultures and their relationship to nature and other species. The colonisation of territories and their organic inhabitants has often led to 'speciescide' as animals are slaughtered to service the consumption of metropolitan human populations. The industrial slaughterhouses of the cities have led some species to the brink of extinction. The permanent human occupation of urban spaces has led to new social relations between humans, species and nature and new forms of entanglement.

Theorising campitude

Camps, domains of temporary habitation and engines of mobility, have been a ubiquitous aspect of the history of human society. They come in many versions from the earliest camps of hunter-gatherers, to the war camps of the Middle Ages, and to the vast encampments of military and colonial forces in war zones. There are camps of refuge and camps of concentration and extinction where temporary inhabitations are moments in the transition of human beings out of this world. From what we can discern about the Neolithic camps of Star Carr to the death camp of Treblinka there is a continuing concern with what humans are, what landscapes they traverse, and the transience of those voyages. Camps have been pirate utopias, the refuges of Maroon ex-slaves, the elegant meeting grounds and fields of medieval kings, camps for discourse, experimentation, and extermination. There have been camps of dissent, revolution and rebellion as in the peasant uprising or the camps of armies waiting to fight Napoleon. The first concentration camp at Norman Cross was designed by the British for Napoleonic troopers many of whom did not survive its 'humane' methods of incarceration. Camps can be of conquest or of flight, in deserts, forests or jungles, even in the heart of cities. Often the camps are camps of imprisonment and captivity. In their many forms there is much to be said about the multivocality of humans themselves and their temporary residences, caravan routes, and paths of flight.

This chapter explores the multivocality of the camps and the kinds of social and human aspirations that support their formation and their persistence. If there is a single univocality across the multiple displays of 'campitude', it is that humans are almost defined by their capacity for and insistence on movement – even when that last camp is part of a mode of death, a destination from which those humans do not emerge, and to which that movement has been compelled and not chosen freely. But that multivocality is also about the social memory of the camp experience and about the ways in which, in the twentieth century, the memory of a specific type of camp would come to define the darkest parts, the 'midnight of the century', in the words of Victor Serge (Weissman 2001: 197). The temporary habitations of those tents and those of the medieval world, and of course even more so those of prehistory, redefine the very essence of traces, for their traces in the material, geological records and in memory are faint and easily dispersed. The deliberate destruction of the traces of the camp at Treblinka are revealed only by a single map, serendipitously drawn at the time by Vasily Grossman, the documenter and archivist of the horrors of both of those terrible camps of Nazism and Stalinism and by the confused, contradictory but essentially truthful memory documents of survivors like Chil Rajchman (Grossman 2011). Further, those political camps echo a world split into abyssal, vast social and ideological encampments – spaces which, even if we are not situated directly in their geographies, we owe emotional and imaginary adherences to and construct solidarities and lineages with.

Camps of extinction, refuge and flight

Primo Levi, in his testimony on and documentation of Auschwitz, recounts his memory of the child Hurbinek. Nobody knew where he had come from or who he was except that he was with them on the trains and on entrance to the camp. All Hurbinek would say was the word/words 'mass-klo/matisklo', a word not one of his fellow encampees could fathom. Neither did they know his ultimate fate (Levi 1979). In many ways this dissolution of identity, a collapse of language, signifies what Giorgio Agamben has called the 'spaces of exception' of the extermination camps where the laws of humans, perhaps even the laws of nature are somehow suspended, spaces which even God seems to have abandoned, or somehow for a moment turned away from.

The camps of that war have also witnessed the most terrifying beauty. Imprisoned in the camp near Görlitz in Silesia, the composer Olivier Messaien wrote the *Quatuor pour la fin du temps* (*Quartet for the End of Time*). Out of Auschwitz came Primo Levi, out of Buchenwald, Ernst Wiechert. There have been camps of quarantine and penal colonies such as the one described by Kafka, with terrifying methods of punishment; the imaginary camps of Narnia and of the Swallows and Amazons. One of the most prescient understandings of imperial camps of contestation has been Julien Gracq's *Les Rivages des Syrtes* (Gracq 1986). How can the imagination not be tempted by travellers in their Romany wagons, or the settlements of the indigenous peoples of the Americas, the camp followers of Brecht and Old Mother Courage and in that same war, of Wallenstein? But for every place of redemption there are enclaves of destruction like Sabra and Shatila and Srebrenica or places where the normal rules of societies are suspended such as the micro-monarchy of the Kingdom of Redonda (Marias 2001). Micro-camps, mountain base-camps, micro-nations, hunting bands, Occupy settlements, camps of desire and drift and social possibility. Some of these camps are Arks, which can save humanity. Some are anti-Arks, which lead it to destruction. Some are extinguished, some become cities. Human beings have long been transhumant, in movement, shepherding flocks and their own peoples. Humans are a specific form of *Neobiota*, organisms which have extended, colonised and imperilled the ecology of the earth. In their camps disaster was waiting for the world and also for them. But for us the camps also raise other possibilities, aspirations and dreams, which can elucidate new manners of being in and upon the world.

Theodor Adorno argued that the 'death' camps, which he himself narrowly escaped and which swallowed so many of his friends and peers, were about a sense of nullity issued to and upon the subjectivity of its subjects, annihilating their autonomy, their self-determination, and sometimes their very being (1974: 16). Adorno was similarly obsessed with houses. Perhaps because of some memory of his own inhabitation in places that were not simply rational, calculable machines to reside in. He remarked upon the very archaic nature of the house and his own enforced condition of emigration with the labour and concentration camps acting as the 'executors' of the 'fate of houses' (1974: 39). These were camps which were invested in technique, replicating the brutality of the guards in the prisoners, making even the 'murdered, murderers' and in which the practice of murder, utterly without pleasure, was even less without measure because of that. Catastrophe, for Adorno, was rehearsed in the camps (1974: 103–104). Even if that catastrophe could be expressed in messianic terms, as in Messaien's apocalyptic *Quartet*, the very idea of the camps as the specific location of the 'burned offering' of the holocaust repels observation. Yet for Blanchot in his writing of the disaster '*It is dark disaster that brings the light*' (1986: 7). That disaster happens in locations which swallow populations and meaning itself 'where all was lost', the disaster is one of disappearance (Blanchot 1986: 27–28).

The camp is the landscape for the disappearance of humans but also the site for the disappearance of the human, as Agamben has said; 'Perhaps concentration and extermination camps are also an experiment of this sort, an extreme and monstrous attempt to decide between the human and the inhuman, which has ended up dragging the very possibility of the distinction to its ruin', (2004: 22). The extermination, argues Agamben, is ultimately comprehensible, even with its 'dark and maimed language', the untranslated word of the child Hurbinek (1999: 37–39). The spaces of exception were spaces of suspension and to which others were indifferent. The practice of extermination was not necessarily unique but the concatenation of murder with location and method was. Agamben states, ‘But this has very much to do with the camps. For what appears in the camps is an extermination for which it may be possible to find precedents, but whose forms make it absolutely senseless’, (1999: 27–28).

The emergence of the contemporary refugee camp, places of refuge and flight, created new kinds of spaces of exception. What begins as informal encampments may develop into new kinds of urban formations and initiate new forms of experience and identity. Michel Agier, perhaps the foremost commentator and analyst of the new camps, stated:

> In the camp, a *place* is formed, and the camp itself becomes the environment where an identity strategy is born, not the contrary, as is often implied. Of course, national and ethnic groups may have existed before the camp, but it is within the camp's space that they are transformed and come up against one another or even mix together; it is there that ethnopolitical forces may come into being with new contours of identity and multiple forms of expression.
>
> (Agier 2011: 284)

But the subjectivities formed within the camps are themselves subject to new systems and totalities. The camps, for Agier, are the ‘left hand of empire’ (2010: 29). They are part of a persisting and durable global system, a powerful apparatus of safety (2010: 30). The new global order of humanity and humanitarianism is a new form of totality for Agier:

> It reaches its fulfilment whenever consensus, the submission of the weak, or the “tolerance” of the dominant erases, stifles, or marginalizes any dissensus that expresses a “disagreement.” Whatever the means through which this consensus is forged, and whatever the shape of the totality represented, there is no longer any excess or outside party whose disruptive voice would threaten the consensus. In the absence of any “parasite” between the whole and the sum of its parts, each part of the whole considers itself to be in an immediate relationship to the whole, sharing the same destiny and coalescing around the same logos. It is a consensual system outside of which there is no remnant.
>
> (Agier 2010: 31)

The disruption and defusing of resistance is part of an imaginary system of ‘kin-based’ conceptions of humanity creating an abstract conception of that planetary humanity at the same time as producing lesser forms of humanity who speak not for themselves and in camps which diminish their very human qualities (2010: 32). In the fictive kin-orders of humanity and its camps there is no residue or alternative:

> At best it is there as an absence, the locus of desire, and the destination of flight. And in both cases, the permanent dualism required by every identification with the social order—which both affords resources and imposes constraints—is the condition for the possibility of subjection without apparent violence.
>
> (Agier 2010: 33)

The geographies and metrics, the calculation and the apparatus of the camp is part of a vast imperial machine which imperils the multiple ways of being of concrete humanity as it submerges their subjectivities. Whether we perceive the camps as new urbanisms (Agier 2002) or as managing 'undesirables' they are seen by Agier as inflictions wielded over subjectivity by imperial regimes.

The mobile subjectivities of human beings are part of the fabrication of the world, enhancements and prosthetics designed to supersede what Bernard Stiegler calls the 'genetic inscription' of those beings (1998: 169). Liberation entails movement and rupture as the human body instrumentalises itself away from inertia. For Stiegler:

> Originary man is practically immobile: finding nourishment under an oak and "finding his bed at the foot of the tree which afforded him a repast," he does not change places, nor move across space, let alone desire to possess the whole of it or even a constantly greater part of it. He is not moved excessively because he is not truly moved, he has no passions: he has natural, vital needs, which he can satisfy in reaping the fruits of nature close at (his) hand—this hand that only grasps without manipulating anything, an organ of grasping rather than of fabrication.
>
> (Stiegler 1998: 114)

The supersession of this inertia and immobility is indelibly linked to technicity and the use of nature (1998: 169). The compulsion towards mobility and the illicit mobility of refugees are in turn mirrored by a whole world of unrestricted movement; of capital, of expertise, and of knowledge. The formation of modernity itself was profoundly related to both forced migration and the vast circulations of experience, resistance, and refuge across the Atlantic abyss. The new utopias of that Black Atlantic were often the consequence of piracy and the 'masterless men' of the oceans, the new nautical proletariat. There were new secessionist communities, new communist colonies, Maroon camps of multi-ethnic revolutionaries at the same time as the prison camps of the Louisiana slave plantations and the cages of the 'wooden worlds' of the ships. The spaces of resistance in the cracks between civilisations created new viral forms of mobile humanity:

> These nomads practice the razzia, they are corsairs, they are viruses; they have both need and desire for TAZs, camps of black tents under the

desert stars, interzones, hidden fortified oases along secret caravan routes, "liberated" bits of jungle and bad-land, no-go areas, black markets, and underground bazaars.

(Bey 2003: 105)

If we think of those routes and interzones as new ways of self-definition and self-determination, then the negative spaces of 'exception' and 'suspension' can also hold other possibilities. New ways of understanding and spatialising the camps of refuge (Ramadan 2010, 2013) can also direct us to other modalities of experience and landscape that can become part of a new utopian 'secessia'.

Camps of secessia

The utopian communities of the past were largely based on the idea of secession and the creation of micro-kingdoms and micro-republics which would act as enclaves within wider geographies, epochal forces, and peoples. Often camps would be created as forms of escape, or of trade, or as deliberate bastions of self-defence. In other ways, as we have seen, we have the temporary inhabitations of exterminatory and labour camps seen by Agamben and others as spaces of exception where the rules of civilisation and morality are suspended. The global policing of refugee camps, often under the aegis of the United Nations or Red Cross, have been for commentators such as Michel Agier part of the same carceral archipelago of capital and empire, dissolving some subjectivities and acting as the engine of others, specifically of the identities of vulnerability and dispossession.

The temporary autonomous zones of Hakim Bey, in secret oases and in the ports of the pirate utopias, have offered new alternatives about the kinds of subjectivities and modes of existence possible within a globalised regime of capital. The spatialisation of the refugee camps, their idiosyncratic geography and identity structures are often appalling and there are many atrocities witnessed to in the camps. They have become liminal spaces for what Engels called 'non-historic peoples' or peoples that history has done a little too much to. They have become spaces that its inhabitants want to escape from or have become new cities like the townships of South Africa. They have developed into an important new modality of inhabitation on the earth, their very uniqueness and temporariness transformed into an enduring facet of how a large part of humanity lives on the planet. Some are largely self-governing, still others testify to contestations between humanitarian and policing powers and the populations they are supposed to advocate for. Indeed, campitude offers a new formulary for inhabitation, new imaginaries of how communities can exist, and novel formulations for what one can experience being part of them. Campitude offers both speculative architectures and new routes for future subjectivities, coalescing less in a 'community' than in a myriad of permeable coalitions, collectivities and self-definitions, changing and transforming, emerging and declining all of the time.

This production of space (where all that is solid melts into air, and all that is air becomes solid, as it were) was expressed over half a century ago in a short work by Ivan Chtcheglov/Gilles Ivain and became a touchstone not just for the developing work of Henri Lefebvre, but also for the emergence of Guy Debord, Raoul Vaneigem and the Situationists with whom Chtcheglov was acquainted. The *Formulary for a New Urbanism* (1953) is reminiscent of many architectural and design manifestos but has a number of distinctive aspects which can illuminate the camps of our contemporary Neobiota and help us think about the experimental architectures of the near-future and the forms of coalescence that will emerge with them. These collectivities have little choice but to be secessionist if we pursue the logics of self-determination and self-emancipation. They can illustrate the adage of another possible world delineated within this one, or what the Situationists thought of as bridgeheads into new worlds. What Chtcheglov calls the 'Quarters' initiates speculative scenarios that (like his Hacienda) can be built. They are built, however, not upon the extinction and suppression of humans and their subjectivity, but on the literal realisation of their imaginaries.

Influenced by his sojourn in the Saint-Germain-des-Prés quarter, Chtcheglov notes the geological and temporal sedimentation of the city, each strata displaying its past:

> Certain shifting angles, certain receding perspectives, allow us to glimpse original conceptions of space, but this vision remains fragmentary. It must be sought in the magical locales of fairy tales and surrealist writings: castles, endless walls, little forgotten bars, mammoth caverns, casino mirrors.
>
> (Chtcheglov 1953: 1).

But these archaic images are just forces of catalysis in the present, the moment between the alive past and the 'already dead future' (1953). The fragmentary strata of the quarters then become the base for the emergence of a new speculative and experimental programme of mobility and of campitudinous transformation:

> Stars and rain can be seen through glass ceilings. The mobile house turns with the sun. Its sliding walls enable vegetation to invade life. Mounted on tracks, it can go down to the sea in the morning and return to the forest in the evening.
>
> (Chtcheglov 1953: 1)

The new experiments in architecture and building are the articulations of new expressions of time and space, multiple iterations of desire – 'Architectural complexes will be modifiable. Their appearance will change totally or partially in accordance with the will of their inhabitants', (1953: 1). Further, this mobile and modifiable experiment with strata, epochs, playfulness, and dreams is itself a new civilisational form expressed in architecture 'On the bases of this mobile civilization, architecture will, at least initially, be a means of experimenting with a thousand ways of modifying life, with a view to an ultimate mythic synthesis', (1953: 1).

And Chtcheglov begins the construction of this new mobile civilisation: The Court of Miracles, the Tower of Nesle, the Sinister Quarter, the Death and the Happy Quarter, spaces entwined with the stories of peoples in movement. For Chtcheglov,

> This new vision of time and space, which will be the theoretical basis of future constructions, is still imprecise and will remain so until experimentation with patterns of behavior has taken place in cities specifically established for this purpose, cities bringing together—in addition to the facilities necessary for basic comfort and security—buildings charged with evocative power, symbolic edifices representing desires, forces and events, past, present and to come. A rational extension of the old religious systems, of old tales, and above all of psychoanalysis, into architectural expression becomes more and more urgent as all the reasons for becoming impassioned disappear...Everyone will, so to speak, live in their own personal "cathedrals." There will be rooms more conducive to dreams than any drug, and houses where one cannot help but love. Others will be irresistibly alluring to travelers...This city could be envisaged in the form of an arbitrary assemblage of castles, grottos, lakes, etc. It would be the baroque stage of urbanism considered as a means of knowledge. But this theoretical phase is already outdated. We know that a modern building could be constructed which would have no resemblance to a medieval castle but which could preserve and enhance the Castle poetic power (by the conservation of a strict minimum of lines, the transposition of certain others, the positioning of openings, the topographical location, etc.)
>
> (Chtcheglov 1953: 1)

The principle of the new mobile civilisation is the drift from one quarter and landscape to another creating disorientation as one tries to navigate and traverse through the multiple camps of the new world.

Of course, this is poetic and speculative and bears the mark both of Surrealism and the Situationists, but it also raises some significant aspects, possibilities and problems in campitude and in using camps as a new, and liberating rather than repressive, way of reformulating the civic spaces of a planetary humanity and its modes of being on the Earth.

The first question concerns the governmentality of spaces. Refugee camps are usually situated in contested or liminal territories. Often the very status of the people or coalescences of peoples are fragile – they may have no access either to civic practices, political status or national identity. They are literally in-between peoples or peoples whose very existence is threatened with annihilation. The governmentality of these spaces are often linked to universalistic discourses of human rights and emancipation, but in concrete terms these essentially non-state spaces have created their own modes of identity and self-governance which in many ways can act as a model for what happens when peoples are brought together in moments and geographies of crisis and catastrophe – in some ways this is the very ethnogenesis of new peoples and new mobile populations. The creation of new forms of subjectivity, sometimes out of horror, is part of new circulations of capital, knowledge and human transfer. The camps elucidate and threaten the idea of borders and boundaries that brought them there in the first place including those within subjectivity itself, its interiors and exteriors, inclusions and exclusions. The lessons that can be drawn from ethnographies of the camps, mapping their geographies, their arts, their new civic practices must also be about the evasion and supersession of the carceral archipelagos of the camps when they are just that.

The extension of global capitalism, new proletarian subjectivities and new technics and machines also offer new modes of self-determination and self-definition. If some camps were designed to determine and define by force against the idea of a universal humanity, creating human and sub-human categories, our camps have to be centred around liberty and expression. This is not to surrender the idea of new epochal, continental and civilisational shifts and counterpose that to the construction of temporary enclaves but to see those temporary inhabitations as marking new maps of those larger shifts; experimenting, speculating and playing with futures. The multiple strata of geographies and histories, of the archaic and of the future, are the resources for that play where quarters are just as much archival retrievals from the past as the experimental architecture of future spaces. Further, they make the idea of 'mobile civilisation' a defining part once more of humanity rather than seeing one part of it as engaging in illicit mobility against and in contestation with static communities of identities rooted in what are perceived as specific and unchanging spaces.

It may be that the temporary inhabitations of the Neolithic revolution and the subsequent projects to dominate nature were the beginning of forms of ultrasociality which would culminate in our potential new Anthropocene epoch – an epoch in which human intervention is signalled within the geological record itself. It may also be that new speculative architectures and urbanisms have to rethink that relationship to geology as well as attend to new forms of sociality which are not quite so inscriptive upon the earth and its records. In fact, it is in those temporary socialities, coming into being and fading, and in those new concatenations of peoples living together, that the real lessons of campitude can be observed. While Messaien was writing his composition about the end of time he looked out from the prisoner camp towards the birds fluttering around the trees, and he listened to and notated their song. Perhaps others still will look into the camps of our future from outside and see the birds singing and winging between those tents and pavilions, and they will step over the border and come in.

This also alerts us to the non-human camps of habitation: those of animals and other species to that of the human. The destructive practices of human beings and their domination of nature may ultimately entail programmes for delineating separate spaces for contesting species.

The meaning of human camps

The construction of stone houses and its relation to prehistory and specifically the Neolithic revolution which itself seemed to transform the human conception of space are one stabilising episode in the long history of the human 'home' (Harkness 2014, Souvatzi 2008). But these home 'camps' are but one iteration or variation of sociality. Reflecting on the concentration of humans into socially occupied homes might see the development of the camp as itself a form of ultra-sociality (Richerson and Boyd 1998). Ultra-sociality is indeed part of the reason for the emergence of the camp, but it is also responsible for the emergence of contestation and warfare between different social groups (and indeed human species) and ultimately the extension of the human Neobiota into cities. The ensuing ecological catastrophe of the human dominion over nature is signalled most forcefully in the suppression of non-human species as the humans invade and use their habitats.

The 'terminal paleolithic' and Mesolithic post-glacial settlement of Britain can help illustrate the emergence of these camps and their significance both functionally and ritually – specifically in the post-glacial settlement of Star Carr (Milner et al. 2013). Not only is the earliest Mesolithic art to be found at Star Carr in the form of a decorated shale pendant (Milner et al. 2016) it is significant for understanding technical economies, tool development (Elliott and Milner 2010) and patterns of settlement (Conneller et al. 2012). More intriguingly it is also significant for understanding the relationship between humans and animals (Little et al. 2016).

Not only were the early Mesolithic sites places of hunting and of ritual, for Chantal Conneller they were localities for corporeal transformation (as all cities would eventually be as vast aesthetic and ritualised extensions of the human body). The original discovery of sets of Red Deer Antler frontlets at Star Carr occasioned commentary, largely by analogy, that they were either or both hunting implements or shamanic head dress. In other words essentially a form of disguise. The original hypothesis of their use as function or ritual has been questioned and developed by Conneller. She argues that this specific locale was a place of repeated visitation in which a variety of tasks took place and materials were left to be re-used. This was unlike other sites around the Vale of Pickering that were abandoned after use. The continuation of the Star Carr site was intimately related to processes of deposition, specifically the Red Deer frontlets, absent from other locales in the vicinity (2004: 42). For functional or ritual purposes the original investigators stressed that the frontlets were related to concealment. As Conneller says: 'Clark saw the antler frontlets as a disguise to conceal the true human body' (2004: 42). The shamanic ritual part of this concealment is questioned by Conneller:

> Rather than being concealed from animals, in this scenario the wearers would be concealed from a human audience. Though in this scenario the person's role or identity may be momentarily altered, it is implicit in this formulation that the body of the person wearing it is not fundamentally or corporeally changed. The mask, here, brings about a "social" or "symbolic" change, not a biological one. In both of Clark's interpretations, the frontlets are therefore regarded as having no fundamental effect upon the body of the wearer, so the only direction for interpretation is determining the context in which they were worn – hunting or ritual. Instead I would like to proceed from a different approach, one that sees the frontlets as revealing, rather than concealing, bodies.
>
> (Conneller 2004: 42)

For Conneller the original vision of disguise and concealment rested upon the binary divisions between humans and animals, humans and things, and the immutability of the body. By using an Amerindian analogy Conneller dispenses with what she sees as the Eurocentric dichotomous 'divisions' of the world (2004: 42). She argues for a 'conspecific' understanding of human and animal relations which stress their cohabitation and mutual interdependence. The 'mutable or relational body' is enhanced and transformed by the material artefacts from other creatures (2004: 43). Following Viveiros de Castro, Chantal Conneller argues that the frontlets are essentially part of an 'assemblage of effects' of the body and that the material artefacts of Star Carr are about ways of relating to and acting in the Mesolithic world (2004: 44). Indeed, because the artefacts were largely the remnants of biological organisms it may be that different kinds of agencies are offered by those body-parts (2004: 45). For Conneller:

> This reassembling of these different objects, after the death of the animal they were once part of, implies that the people who occupied Star Carr retained a conception of the "animalness" of both artefacts. Rather than viewing these objects as passive recipients of meaning, as retaining some symbolic sense of "animal", these artefacts can be seen as incorporating elements of their original animal agency or, more particularly, as animal effects. So taking on, for instance, barbed points made of red deer antler would allow people to harness antler effects in order to enter into particular relationships with the world and in doing so to incorporate new perspectives on it.
>
> (Conneller 2004: 47)

The use then of the animal bodies as prosthetic extensions of the human body discounts the idea of a biologically bounded immutable entity which is stable and unchanging. Decoration and ornamentation increase the powers and effects of the individual human and by extension the spaces and socialities that they inhabit (2004: 47). This extension then problematises the seams and borders between the animal body and the human body with the human altered and extended by their relation and adoption of animality. The bodily boundary and perspectives on the world are then 'deer-effects' (2004: 47–48).

> Becoming animal is not about moving between different bodies. Though the animal is not literal, the transformation is. Parts of human bodies connect with parts of animal bodies to produce a new assemblage of bodily effects, one that is something else entirely, not human, not dog, that relates to the world in a new way. Just as the animal part transforms the human body, so the conjunction with human parts transforms the animal.
>
> (Conneller 2004: 50)

The sustained deposition at Star Carr, its repetition and reiteration is significant locationally (2004: 51). It is a site of stability, inhabitation, consumption and a location which brings together animals and humans in new variations.

But how had the post-glacial recolonisation of these spaces taken place and what were the mobility and settlement strategies of those new human groups and their camps? It is worth returning to Conneller at Star Carr because in her conclusion she illustrates something of some significance about the cosmology and epistemology of these new settlements. There is a methodological problem, Conneller argues, of distinguishing between differing types of settlement but there is also the problem of what we might call the 'boundlessness' of mobile, migratory populations which were initially sparse and did not exist in any kind of circumscribed or restricted space. For Conneller: 'People encountering vast, empty, unknown places may act in ways that cannot be imagined with reference to more recent hunter-gatherers', (2007: 216). Using analogies from contemporary hunter-gatherer populations who are 'bounded' may then be misleading. Conneller argues that:

> Detailed examination, therefore, of even a single archaeological site reveals that the mechanisms of recolonization may be more complex than previously envisioned. The imposition of a blanket model of pioneer/residential settlement precludes an examination of the varied technical and settlement strategies through which people might be able to move into new landscapes. In addition to the movement of small hunting parties from adjacent 'settled' areas, archaeologists should be open to the possibility of highly mobile social groups practising forms of mobility for which there are no recent analogues.
>
> (Conneller 2007: 217)

The technical economies of the Final Palaeolithic moving into the Terminal Palaeolithic and Mesolithic are difficult to discern (2007: 227). Indeed, as

Straus has argued the meaning of transition between these epochs is fraught with misunderstanding (Straus 2009). An understanding of the technical tool developments, the origins of materials and their transportation is then one of the few ways we can distinguish human cultural movement. For Conneller:

> The relationship between technical economies and mobility appears to have been configured rather differently in the Terminal Palaeolithic. At all the sites examined large quantities of material appear to have been imported, often from non-local sources on the North Sea Plain. In contrast to the finished tools of the Final Palaeolithic, raw material appears to have been preferentially imported in the form of pre-formed cores and stocks of blades, with more occasional instances of finished tools or tested nodules. Material was also removed from these sites, sometimes in the form of partially worked cores, but there seems a definite concern to retain and replenish stocks of blades for future use. A stock of blade blanks, as well as being of use in future campsites, would have been a flexible way of engaging with a range of situations in the broader landscape. In contrast to the Final Palaeolithic, the impression is of people who moved frequently and flexibly but possibly – burdened by greater quantities of raw material – over shorter distances.
>
> (Conneller 2007: 232)

And this returns us to the wonderful, concluding insight of Conneller: 'While at the end of the Upper Palaeolithic people brought their world with them, during the Mesolithic the world was revealed and maintained through a constant re-engagement with particular places', (2007: 234–235).

The transport of one's cosmological and epistemological universe with the corporeal beings and tools of human beings entails the unbounded universe of deer tracks and forests. The sparsity of population allowing for the mobile extension of humanity without contestation over locations, spaces and the 'earth'. The rootedness to specific locations that emerges in the Mesolithic and is fully confirmed with the 'empire of neolithia' secures both an agricultural and a cosmological revolution which stabilises populations in locations or a set of locations (Hudson 2015b). The universe becomes bounded by walls and eventually locations are secured to ancestries and genealogies of power with the burial of ancestors in specific locations to demarcate ownership. This created the tribe in which power was divided and demarcated across populations living in separate bounded spaces with invisible borders imposed upon the landscape often with an implied relationship to natural features such as hills or rivers.

It created new relationships of interiority (what I am, who I share my identity with, how do I differ from other humans, from animals, from rocks) and exteriority (who are the strangers, where do they live, what is their identity as different to mine). The locatedness in specific plots of earth implied both a ritual and functional/economic relationship with the landscape: its borders were crossed as new paths of connection for ritual and functional reasons connected up different tribal structures and eventually large confederations of peoples like the Brigantes. These were still transhumant peoples but they were not constantly mobile across different territories but moved seasonally subject to the powers of their cattle, weather and crops. Their relationship with the cosmos changed with gods and spirits inhabiting specific locations of rocks and streams, the petrified memory of ancestry, war or grief enshrined within specific landscape forms and burial places. The spaces of the human and the non-human were transformed with caves perceived as the liminal locations between the living and the dead (Hudson 2017). The wilderness and the space of humans became conceptual ways of dividing the earth with new forms of aesthetics which represented the ways in which humans related themselves to animals and their powers. The subjection of animal life to mercantilism and commodification would be one part of the propulsive mechanisms of early capitalism at the same time as relations of labour changed with the development of machinery. Ultimately the agonistic contestation between demos and power, between master and proletariat would shape its own modernistic cosmology: the looms of the masters replacing the warp and weft of the primitive labourer. The class structures of prehistory and antiquity transformed time and time again as new forces and relations emerged which, whilst preserving archaic social forms into new epochs, utterly transformed the human relationship to nature and humanity's own nature.

Examining briefly one of those early cities can illustrate something important about conquest, nature and the human space: the city of Troy (see Hudson 2016a for a more sustained account of this motif). Austin, in an early analysis of the Troy motif in Virgil, notes the relationship between corporeal transformation of the 'toy' horse as a prosthetic tool of conquest and invasion: 'Virgil knew this very problem; and he knew that in handling the tangles of myth that invested the Tale of Troy, diversely treated before him by many writers, he must allow nothing to distract attention from the single overwhelming fact, *Troy is overcome, And Priam dead*', (1959: 16). What is fascinating about the wooden horse motif is the use of the model to destroy a city, its stomach containing the conquerors who will murder the children of Troy and raze the city to the ground:

> First the Horse appears; then Laocoon gives his warning, Sinon tells his tale, Laocoon and his sons perish; the Horse is taken into Troy, and its menace is fulfilled. The last name in the list of Greeks who come swarming out of it is that of its builder, and the pattern is finished. We are conscious of the Horse's presence throughout the scene.
>
> (Austin 1959: 16)

The horse is both a symbol of peace and the tool which will bring to birth the destruction of a people and the rebirth of the new Achaean people: their empire and their new cosmology resting as it does upon the relegation and the destruction of another people and another set of gods. But at this moment of birth: 'The men in its vast depths have as yet no names; they are dark, menacing shapes, waiting in its womb', (1959: 17). The horse, far from birthing a new moment of peace and reconciliation 'is a practical and deadly instrument of war, the doom of Troy', (1959: 17). The symbolisation of the wooden horse taken within the city walls only to destroy that city is ubiquitous:

> In modern times it has been held to be a siege-tower of Assyrian type, a fertility-symbol of "hobby-horse" type, a magic antidote to the magic walls of Troy, a theriomorphic manifestation of Poseidon, and a rationalizing of a miraculously parthenogenetic mare. But Virgil seems not to have been aware of any of these possibilities: a Horse beside the river's brim a Wooden Horse it was to him, and it was nothing more.
>
> (Austin 1959: 23)

Indeed, for Virgil, the horse is simply what it is: an instrument, a tool, a prosthetic. It is neither metaphor nor mythological representation of something else. It is a model simply of a horse, albeit one pregnant with the dark forces of destruction who will leap down from its belly. Marx himself was obsessed with the classical and specifically notes the importance of the epic of Troy for understanding the profoundly different cosmologies of the ancient world and the capitalist cosmos. Marx argues that the epic of the classical world cannot exist in the same form in the present work. The ideas of nature and the social relations of Troy and the Greek imaginations are not compatible with the 'with self-acting mule spindles and railways and locomotives and electrical telegraphs?' (1973: 110). The Iliad of Homer and specifically the drama of Achilles end with the emergence of the 'machine' and 'can never return' (1973: 111). The emergence of the organic/mechanical horse of Troy is pregnant with the social formations that will follow in the wake of the conquest of Troy. Indeed, the horse as a machine signals the beginning of the new world of automata and prosthetics that the Greek world was pregnant with. Marx asks what are then the epics of Capital? (1973: 111) Indeed, they can be seen in the epics of machinic destruction and power; in Moby-Dick and in the power of Marx's writing on the machine itself.

The metropolis and modernity

> Every day he hears about the metropolis. The name of the city changes. It is all cities, overlaying one another and becoming a city that exists nowhere but which continually transmits promises. These promises are not transmitted by any single means. They are implicit in the accounts of those who have already been to a city. They are transmitted by machinery, by cars, tractors, tin-openers, electric drills, saws. By ready-made clothes. By the planes which fly across the sky. By the nearest main road. By tourist coaches. By a wrist watch. They are there on the radio. In the

news. In the music. In the manufacture of the radio itself. Only by going to this city can the meaning of all the promises be realized. They have in common a quality of openness.

(Berger 1975: 23)

John Berger sees here the city is an imperial force, a ghetto, a refugee camp, and a hybrid social formation (see Castree 2000a, Cole 2003, Edkins 2000, Swyngedouw 1996 respectively). But Berger also sees the city as entity which elevates and creates a quality of openness. It is an enclave of the human beings amongst nature and animals, serviced by and subjugating nature and animals. Georg Simmel expresses the condition of life in the metropolis:

> The deepest problems of modern life derive from the claim of the individual to preserve the autonomy and individuality of his existence in the face of overwhelming social forces, of historical heritage, of external culture, and of the technique of life. The fight with nature which primitive man has to wage for his *bodily* existence attains in this modern form its latest transformation.
>
> (Simmel 1969: 47)

The emergence of social science as an independent discipline largely rested on the social problem of the city and the problem of the organic and mechanical analogies for human beings and for whole social formations: specifically for Simmel in the subjection of the individual human being to becoming a cog in the 'social technological mechanism' (1969: 47). The 'autonomy' of the individual is in contestation, as Simmel says with four forces: the social, the historical, the cultural and the mechanical. She is arrayed against 'nature', as a technical struggle, in order to reproduce her 'bodily' existence. Only the subjugation and subjection of nature can allow the individual body to reproduce and to maintain its resilience as a natural, organic entity in the midst of the city as a machine. Her very bodily autonomy enmeshed in social and urban formations is premised upon natural destruction. The autonomy not only extends the social powers of the body beyond the biological and physical organism it reproduces itself across geography and history. For Simmel:

> This is the counterpart to the independence, and it is the price the individual pays for the independence, which he enjoys in the metropolis. The most significant characteristic of the metropolis is this functional extension beyond its physical boundaries. And this efficacy reacts in turn and gives weight, importance, and responsibility to metropolitan life. Man does not end with the limits of his body or the area comprising his immediate activity. Rather is the range of the person constituted by the sum of effects emanating from him temporally and spatially. In the same way, a city consists of its total effects which extend beyond its immediate

> confines. Only this range is the city's actual extent in which its existence is expressed.
>
> (Simmel 1969: 56)

It is indeed, like the deer 'masks' of Star Carr, a corporeal transformation in which the 'effects' of the body are profoundly extended across time and space moving beyond function and ritual into a revolutionary transformation. The city becomes a qualitatively different organism from just its physical presence. Marx says something similar:

> With its coming-together in the city, the commune possesses an economic existence as such; the city's mere *presence*, as such, distinguishes it from a mere multiplicity of independent houses. The whole, here, consists not merely of its parts. It is a kind of independent organism.
>
> (Marx 1973: 483).

Cornelius Castoriadis, in his discussion of the urbanism of Yves Barel, takes the medieval commune as his starting point for understanding the social imaginaries and effects (the magma) of the city (Castoriadis 2003). The development of complexity indeed offers new ways of thinking about the autonomy of human beings and their communes and collectives if only the city can be transformed in order to initiate new ways of thinking about human freedom, even if that very freedom is based upon the suppression of nature. Engels talks about the 'colossal centralisation' of human beings, machines and engines of transport (1969: 55). For Engels, the city is an agent of universal transformation:

> This colossal centralisation, this heaping together of two and a half millions of human beings at one point, has multiplied the power of this two and a half millions a hundredfold; has raised London to the commercial capital of the world, created the giant docks and assembled the thousand vessels that continually cover the Thames. I know nothing more imposing than the view which the Thames offers during the ascent from the sea to London Bridge. The masses of buildings, the wharves on both sides, especially from Woolwich upwards, the countless ships along both shores, crowding ever closer and closer together, until, at last, only a narrow passage remains in the middle of the river, a passage through which hundreds of steamers shoot by one another; all this is so vast, so impressive, that a man cannot collect himself, but is lost in the marvel of England's greatness before he sets foot upon English soil.
>
> (Engels 1969: 55)

Indeed, it was impossible to even set foot upon 'English soil', earth and landscapes utterly diminished and covered over by the pavements and machines of the urban metropolis. Yet these urban spaces, at once offering autonomy can also be understood as similar to the camps of extermination spaces in which humans disappear and indeed the very idea of the human might disappear, destroying the distinction between the human and the inhuman (Agamben 2004: 22). But perhaps not in the way that Agamben means in his analysis of the 'camps' but in the dissolution of the human when faced by its own prosthetics of production and destruction. On arguing for a pause in the development of Soviet city construction and the vast social engineering projects of engineers like Tatlin (who would himself become obsessed by combining human and animal into new flying machines) Leon Trotsky pays tribute to their ambition:

> The originators of gigantic projects, men like Tatlin, are given involuntarily a respite for more thought, revision and for radical re-examination. But one must not imagine that we are planning to repair old pavements and houses for decades to come. In this process, as in all other processes, there are periods of repair, of slow preparation and accumulation of forces, and periods of rapid development. As soon as a surplus will come after the most urgent and acute needs of life are covered, the Soviet state will take up the problem of gigantic constructions that will suitably express the monumental spirit of our epoch. Tatlin is undoubtedly right in discarding from his project national styles, allegorical sculpture, modelled monograms, flourishes and tails, and attempting to subordinate the entire design to a correct constructive use of material. This has been the way that machines, bridges and covered markets have been built, for a long time. But Tatlin has still to prove that he is right in what seems to be his own personal invention, a rotating cube, a pyramid and a cylinder all of glass.
> (Trotsky 1991: 274–275)

As Trotsky was looking towards the city and indeed the new machinic humans of the future he was waging a war not just against the white armies of reaction but also against nature itself with Lenin's injunction for the electrification of the USSR. The abandonment of the mass industrial soviet project and its new elaboration of prosthetic human beings in ecological catastrophe and pollution paralleled similar developments in the global south, east and west. The abandonment of global generic forms of human progress and the 'human movement' began to reassert the vengeance of nature against their human contestants.

When H.P. Lovecraft began to develop his Chthulu 'mythos' and cosmology he was an avid reader of Scoresby's accounts of the Arctic. His vision of a monstrous Antarctica of the south paralleled the whaler's vision of the north and Lovecraft was obsessed with the mirages that he read about in Scoresby's account of the icy north. These would be replicated in the horrifying mirage of a vast Antarctic anti-human city in 'At the Mountains of Madness' (2007). The story begins with an account of the machinery and technology of a geological expedition to Antarctica with two ex-whaling ships, the *Arkham* and the *Miskatonic* captained respectively by Douglas and Thorfinnsen (2007: 278–280). The narrator encounters early in the expedition the same phenomenon that Lovecraft had discovered in Scoresby: 'On many occasions the curious atmospheric effects enchanted me vastly; these including a striking vivid mirage – the first I had ever seen – in which distant bergs became the battlements of unimaginable cosmic castles', (2007: 280). This is of course a fabulated, fictitious vision but it hints at horrors to come as the expedition begin to encounter remnants of an 'unknown and radically unclassifiable organism of considerably advanced evolution' (2007: 285). The destruction of the laboratory tents leads the narrator into the interior, and into the city of the 'old ones', the creatures from outer space who had existed in this place for millennia before the advent of humans and simians. The account is of a city of monstrous proportions:

> The effect was that of a Cyclopean city of no architecture known to man or to human imagination, with vast aggregations of night-black masonry embodying monstrous perversions of geometrical laws. There were truncated cones, sometimes terraced or fluted, surmounted by tall cylindrical shafts here and there bulbously enlarged and often capped with tiers of thinnish scalloped disks; and strange beetling, table-like constructions suggesting piles of multitudinous rectangular slabs or circular plates or five-pointed stars with each one overlapping the one beneath. There were composite cones and pyramids either alone or surmounting cylinders or cubes or flatter truncated cones and pyramids, and occasional needle-like spires in curious clusters of five. All of these febrile structures seemed knit together by tubular bridges crossing from one to the other at various dizzying heights, and the implied scale of the whole was terrifying and oppressive in its sheer gigantism. The general type of mirage was not unlike some of the wilder forms observed and drawn by the arctic whaler Scoresby in 1820, but at this time and place, with those dark, unknown mountain peaks soaring stupendously ahead, that anomalous elder-world discovery in our minds, and the pall of probable disaster enveloping the greater part of our expedition, we all seemed to find in it a taint of latent malignity and infinitely evil portent.
>
> (Lovecraft 2007: 302)

The urban ethnographies of the 'old ones' that the narrator engages in as he traverses the city is itself an anatomy and anthropology of a profoundly non-human entity (2007: 332–333). Like the beginning of the human camps at Star Carr and their corporeal transformation as they engage with other species, so the Antarctic city returns us to other transformations, epochs and corporealities and perhaps the destruction of the human species itself.

5 Locomotive cultures

The extension of human power over geology and geography is discussed in this chapter, particularly in relation to the 'locomotive' machinic cultures of the earth. The idea of bounded territories has been challenged by a global, mobile culture – specifically and initially that of the train-engine. Intimately related to empire and colony the train was a machine for the transportation of populations and labour, of commodities and raw materials and of armies. This chapter looks at the cultural history of the development of locomotive power and its relationship to animals and the natural world that the 'engine' traverses. However, first and foremost, the development of the prosthetic tool entailed the biological development of the hand. As we have seen, this was noted by Engels who saw the hand as central in the transition from 'ape to man'. In fact, as Marx makes clear in the *Grundrisse*, the hand is the foremost power of humanisation – an idea that has been taken forward by structural anthropologists like André Leroi-Gourhan. Indeed, it is the work of Leroi-Gourhan that, in pointing to the hand and the extended tools of motive-power and locomotive-power, sustains our understanding of machine-construction as a process of exteriorisation: that process by which the mental potentialities of the brain become materialised in production. This chapter, then, examines the role of the hand in the production of technics and then moves on to try and understand the contribution of Leroi-Gourhan to the history of the machine. It then addresses the dual nature of the steam-engine and the locomotive as both dream and nightmare before addressing what Paul Virilio has called 'metabolic vehicles'. It concludes by examining those rail routes that transported other species into the slaughterhouses of modernity.

Railways have been, in their various representations, synonymous with liberation, domination, catastrophe and accumulation. However, the fundamental feature of their material manifestation is their mobility. They also signify a profound relationship between the individual human hand and the vast systems of transportation that are prosthetic extensions of that hand. Hannam, Sheller and Urry have noted that complex mobility systems are often specialised and arcane (2006: 4–5) but they also stress the central relationship between mobile systems and their moorings and immobile parts (2006: 15–16). The infrastructure of these vast, almost accidentally produced, mobile systems depends upon a series of static parts that are immobile; stations, depots, tracks. So much so that in terms of governance and political economy, the static trackways and infrastructure are often separated from the engines that plough their ways across those immobile structures. Like the vast machines of the industrial factory system humans have become appendages of the machines, their hands simply used as sentient cogs. This is within what is for all intents and purposes an almost fully-automated system of engines and fixed structures.

These systems are complex coalescences between organic and mechanic elements. This does not mean that they are coalitions or alliances. This would intimate some sense of equality between sentient and non-sentient elements whereas in fact the human is subordinated to the vast system of transportation. Where human interventions do occur within the system they often take the form of disruption or suicide which is often repaired by other sentient elements working for the automated monster of that system. These temporary aberrations are simply parentheses in the history of the system: momentary suspensions of its activity in a specific space and time. Yet these catastrophes and accidents often became the stock tropes of Victorian literary fictions hinting at both the fascination and fear of the machine and of its disruption.

John Tresch, in his survey of 'mechanical romanticism' (2012) has argued that the way we use machines is intimately related to how we understand nature. The knowledge and observation of nature is, for Tresch, about specific theories of nature and theories of 'knowing'. In the post-Napoleonic period, as new types of machines emerged, like the steam-engine, so did a new conception of nature emerge. Tresch calls these 'romantic machines' – ones in which both animated living elements were combined with inanimate mechanical elements. Unlike the mechanical analogy of the 'classical machine', human sentience and consciousness was central to the flexible and proper functioning of these new machinic structures (2012: xi). Bernard Stiegler has, however, argued that these labouring machines do not make capitalism work as such. Capitalism is the production and the extension of the libido, of the dynamics of desire (2014: 41–42). In this sense, what Marc Bloch calls the 'motive-force' of capital accumulation is for Stiegler simply a kind of armed desire (2014: 42). Further, it could be argued that it is directly carnality and specifically 'carnism' that is the desiring motive-force of the engines of accumulation (Nibert 2014). This is particularly true of the vast networks of rail that transported cattle into the slaughterhouses of Chicago, that allowed the migration of human appendages of labour across continents, and compelled the cattle-trucks to make their way to the death camps with human rather than animal 'products'. This is not to make an equivalence between the attempted destruction of the Jewish people and animals but to see the movement of animals as prefiguring the Holocaust.

The desire for destruction is at once consumption – in the latter case whole peoples were consumed by the 'ovens'. Indeed, the experimental laboratories of the cattle-trucks and the transportation of cattle, including the transportation of the products of the Cetacean slaughter from the railyards of Europe, directly prefigured the human holocaust, mapped its routes and destinations, tested its logics and logistics. The extension of the human hand therefore made possible the subjugation of nature, including human biological nature.

Motive-power, technics and the hand

Edward Baines, an early textile historian, bears witness to the arrival of the steam-engine:

> All are moving at once – the operations chasing each other; and all derive their motion from the mighty engine, which, firmly seated in the lower part of the building and constantly fed with water and fuel, toils through the day with the strength of perhaps a hundred horses. Men, in the meanwhile, have merely to attend on this wonderful mechanism, to supply it with work, to oil its joints, and to check its slight and infrequent irregularities…
>
> (Jones 1973: 47)

The social struggles of the weavers across the steam looms and the destruction of the mills were an attempt to decelerate industrial capitalism and their own subordination as appendages in its machinic structures. Marx had witnessed, as we have already noted, the human becoming an appendage and as a fragmentary being in *Capital* but he was contemptuous of attempts to restore pre-capitalist economic and social formations by individualised labour. In the *Grundrisse* Marx saw that the division between nature and the machines opened up potentially endless elaborations of 'human nature' if only labour could achieve mastery over the machines and thereby the collective control of nature (Geras 1983: 80).

The whole point of writing the notebooks of the *Grundrisse* was to lay bare the object of material production and explore not just the human as a social animal or 'animal of the polity', but to examine social forms against 'wilderness' (Marx 1973: 83–85). For Marx (and Engels) the congealed evolutionary history of the hand was the essential clue to the enigma of material production and the transitions that production had made since prehistory. For Marx 'No production possible without an instrument of production, even if this instrument is the hand. No production without stored-up, past labour, even if it is only the facility gathered together and concentrated in the hand of the savage by repeated practice', (Marx 1973: 85–86). This phenomenology of the hand is perhaps the most original contribution of the *Grundrisse* and is enmeshed in both production and consumption. The hand of the 'savage' can tear and feed meat into the mouth but production refines this consumption, makes it mannered and civilised. The product of dead flesh is profoundly different from the product of prepared meat even as the object literally de-composes as it is eaten thereby recomposing and reproducing the organic being that consumes it. As Marx says, this process of desire, appetite, gratification, consumption, is the subjectification of the material object to ourselves as humans just as we objectify ourselves in its production (1973: 88–89). Production then, is immediately consumption with the taking in of food as the production of one's own body (1973: 90). Unlike a directly 'natural' product, the food only becomes a product through consumption and only by the decomposing of the product does the consuming take place (1973: 91). The 'manner' of consumption is of course socially determined by differing modes of savagery or civilisation. The dissolution and decomposition of the entity (product, consumable) dissolves it into our own metabolism as its own is extinguished.

There are of course different modes of exchange and distribution of products but the essential point holds, that it is the process of hand to mouth and the social determination of the relationships around it that compels both production and consumption. But there is also something more profound about that act of consumption. For Marx:

> Bourgeois society is the most developed and the most complex historic organization of production. The categories which express its relations, the comprehension of its structure, thereby also allows insights into the structure and the relations of production of all the vanished social formations out of whose ruins and elements it built itself up, whose partly still unconquered remnants are carried along within it, whose mere nuances have developed explicit significance within it, etc.
>
> (Marx 1973: 105)

This points us to the archaic elements that are still present within our social formations, composed as they are of the disjecta membra or detritus of the past. But consumption also propels the continuing passage of archaic organic beings – essentially ourselves – who in both our bodies and our consciousness constantly recompose and reproduce the archaic. Indeed, one feature of some types of consumption is not just the dissolution within our own metabolisms of expiring beings but of entities long dead, rare fruits and wines as they are. The organic development of the hand for Marx is propelled by the activity and the labour of eating and of production more generally but also nourished through the subordination of other organic beings through consuming them. The hand is 'capital', accrued and accumulated, in its starkest form (1973: 257).

In his analysis of Marxist aesthetics, Ernst Fischer elaborated the theory and phenomenology of the 'hand'. Fischer sees the hand as essentially an organ of experimentation, and although he recognises the importance of muscular structure and vision, the decisive productive organ was the hand:

> The pre-human being which developed into man was capable of such development because it had a special organ, the hand, with which it could grasp and hold objects. The hand is the essential organ of culture,

> the initiator of humanization. This does not mean that it was the hand alone that made man: nature, and particularly organic nature, does not allow of such simple and one-sided sequences of cause and effect.
>
> (Fischer 1963: 16)

When examining the hand Fischer notes again what he calls the 'mediated metabolism' of the human-nature relation (1963: 18).

> Every biological organism is in a state of metabolism with the surrounding world – it continually gives and takes something to and from that world. But this taking is always done directly, without an intermediary. Only human work is *mediated metabolism*.
>
> (Fischer 1963: 18)

The beginning of the prosthetic extension of the human hand creates a situation in which the instrument of labour is outside the organic structure of the human body and therefore cannot be injured but may be replaced, modified and redeveloped. Primitive tools can be replaced by more developed ones. Rather than the human being adapting themselves to nature, this non-organic instrument is adapted to the needs and desires of the human being and is the beginning of the 'mastery' and the subjugation of nature through technics. Further,

> And out of the increasing usefulness of his tools, out of their increasingly specific character, out of their increasingly successful adaptation to the human hand and the laws of nature, out of their increased *humanization*, objects were created which could not be found in nature. More and more the tool lost its resemblance to any natural object.
>
> (Fischer 1963: 22)

Humans have then, for Fischer a 'double nature'. They have emerged from nature and are at least partly natural beings but they have created what Fischer calls a 'counter-nature' or 'super-nature' (1963: 32). Through human labour each object is fashioned, at first a semblance or model of something and then transformed into a prosthetic tool. This happens at the same time as material objects are turned into signs and names and concepts and human 'animality' begins to disappear (1963: 33). The system of propulsion transformed the earth as well as transformed the nature of human beings themselves. This 'counter-nature' is more fully developed by André Leroi-Gourhan and his rendition of the nature of the hand.

Technics and Leroi-Gourhan

André Leroi-Gourhan's work has influenced a range of subsequent commentators including John Berger, David Lewis-Williams and Bernard Stiegler. His structural anthropology and specifically his structural analyses of Palaeolithic cave art, have been central to understanding the emerging concept of the 'human' (Leakey 1994: 110–111, Lewis-Williams 2002: 59–65). Cornelius Castoriadis, in his own analysis of 'technique' has pointed to the work of Leroi-Gourhan on the 'technical ensemble'. If the tool is, as we shall see, the exteriorisation of the interior being of humans then the material object is not a pure 'instrument' per se but enmeshed in a whole series of significations and social imaginaries (1984: 245). For Castoriadis:

> The gulf which separates the necessities of man as biological species from the needs of man as historical being is dug by the human imaginary, but the pick which digs it is technique. Yet this image is still inadequate, for here once again technique taken *in toto* is something more than a simple instrument, and its specificity always decisively co-determines what is dug: historic need cannot be defined without reference to its object. Industry is not just 'the open book of the essential powers of man', it is also the endlessly continuing text of the impossible translation of desire into realisable intention.
>
> (Castoriadis 1984: 241)

For Castoriadis there is no limitation and no foresight to what the human being is capable of. The social imaginaries of human beings, that mark them off from *Homo sapiens* as biological organisms, allow for the elaboration of a ceaseless relationship between desire and object. Desire is not simply a biological condition or imperative to satisfy an appetite, to gratify a physical need, but a desire which is constantly reshaped by the imaginations of the human collective or its multitude of collectivities each with their own social imaginaries. But for Castoriadis and developing on the structural aspects of Leroi-Gourhan, it is the material object itself that makes manifest the imaginary,

> this object is itself a product, and its genesis therefore brings into play the whole social existence of the collectivity which gives birth to it – not just its 'mental aptitudes', but its organisation of the world and the particular bias by which this is characterised. It is not simply that each culture (or class of cultures) has its own 'style' of inventions and artefacts, which more or less corresponds to what Leroi-Gourhan calls the 'technical group'; the point is that the technical ensemble gives concrete expression to a grasp of the world.
>
> (Castoriadis 1984: 245)

For Castoriadis the culture does not create the taxonomies of tools and neither can we interpret the society from its taxonomic tool structure. The technical tool and the imaginary co-exist and are entangled. Indeed, the social imaginary itself is deeply implicated with the notion of the hand and its extensions.

Bernard Stiegler develops this notion of the extension of the hand that we find in Leroi-Gourhan into the whole phenomenology of the social and the natural world:

> Prostheticity is the foundation of human aesthetics as that pros-thesics that can only take shape [*prendre corps*, come into being, materialise] with the conquest of the upright stance, of erect posture, the inaugural moment of a process through which the hand, abandoning its locomotive function, invents a fabricative function. The de-functionalisation of the paw, which thereby becomes the hand or the foot, is the very opening of technicity, and constitutes a re-functionalisation (a functional re-equilibration or readjustment, as André Leroi-Gourhan says): the hand is then the producer of signs, objects, artefacts, prostheses, and works. And the foot starts to dance. This working hand opens a world [*cette main ouvrière ouvre un monde*]
>
> (Stiegler 2010: 152–153).

The shift from the locomotive hand to the fabricative hand literally 'makes' the world – not just of the object but the conceptual ideological systems of the human. The hand both fabricates and 'signs' at the same time as it inscribes the workings of the human imagination onto the fabric of the earth. This inscription is a writing of the 'digits' creating the technical, third memory of the machine onto and into nature (See Olsen et al. 2012: 105). Understanding the insights of Leroi-Gourhan in more depth can help illustrate the locomotive and fabricative phenomenology of the human hand.

In his examination of the structural languages of tools in *Gesture and Speech* (1993) Leroi-Gourhan elaborates a new anthropology of human making (see also Ingold 2013 for a magnificent analysis of the anthropology of making). Leroi-Gourhan's work is diverse and complex making it difficult to present a comprehensive analysis even of this single work. As an archaeologist and anthropologist, he was obsessed with the entanglement and contact between humans and their technics. He begins his elaboration of their relationship by examining the question of 'distancing'. This is the process by which human beings separate and dislocate themselves from other things around them and from their own biological being and condition. The detachment is manifested in the separation of hand and prosthetic tool but also in terms of separating out a material object from our concept of it. Further, in the development of those distancing reflexes, humans separate themselves from other zoological species. It places the human outside the systems of taxonomy that we use for other animals. Perhaps critically, the distancing process is one in which things external to us become depositories of memory, transferring memory into other objects and social organisms. Technical memory creates a schism between the organic human memory which is extinguished and the relics and remnants that humans put into their technics, their products and their art (Leroi-Gourhan 1993: 235). This process of exteriorisation maintains individual human memory structures potentially across vast epochal time scales. The inscriptions of Upper Palaeolithic cave art that Leroi-Gourhan mapped and structured clearly indicate that process of exteriorisation, memory and sustainability. The meaning of these memory forms, hypotheses that Leroi-Gourhan obsessively pursued, would however be more enigmatic (Lewis-Williams 2002: 60–61 and see also Derrida 1976 on Leroi-Gourhan and inscription).

The locomotive and fabricative technologies of the hand were the single most important thing to separate and demarcate the human from its own animality. The hand, however, is in intricate relationships with the development of *mind*. Leroi-Gourhan argues that the brain has an imaginative capacity to think everything even as it is empty at birth (2003: 228). This gives the human mind an almost infinite capacity for development beyond its organic substrate. But this mind is, more than anything, a collective social mind. Its emptiness is filled with the social imaginary. The mechanisms of social memory are, for Leroi-Gourhan, about 'preservation' in order to repeat and ritualise necessary social activities like building, hunting, growing. Proverbs and recipes are stored in the human mind in oral communities but mechanical storage of memory becomes more necessary through inscriptive devices such as writing and language (2003: 234). The animal is compelled by its genetic memory and the human by the 'ethnic' capital of its historical community and its traditions (2003: 234).

The reliquaries and depositories of the social memory are themselves the product of the hand and of the 'technical gesticulation' of grasping (2003: 238). For Leroi-Gourhan different modes of grasping intimate different forms of prosthetic machines:

> Action peculiar to the hand consists in the potentially wounding effect of fingernails, in grasping operations involving the fingers and the palm (digitopalmar prehension), and in grasping between the fingers (interdigital prehension). Since movements of transmission or rotation determine both our manner of holding a hand tool and the impetus we apply to it, a fourth term descriptive of the leverage exercised by the forearm and the arm is needed in order to analyze human gestural behavior in the technical field.
>
> (Leroi-Gourhan 2003: 238)

Devices were developed by humans for 'grasping, transporting, and positioning objects' (2003: 243). The crane and the pulley block as ancient technologies were simply intuitive extensions of the motive-power of the hand. It would exteriorise the will of the brain into a tool (2003: 243). But the gradual extension of the machines would distance labour from the human fingers more and more. Beginning in the Neolithic the 'freeing' of the fingers would introduce new automotive processes that would gradually separate, indeed alienate, labour from its material objects and production. As Leroi-Gourhan says:

> Not until the nineteenth century did the introduction of a punched-card system raise mechanical weaving to the level of handling skill of which the bare hand had been capable from the start. In both cases the development is the same: In the first stage, the hand can perform actions that are limited in terms of force or speed but infinitely diverse; at a later stage, that of the pulley block or the weaving loom, a single action of the hand is isolated and transferred to the machine; in the third stage, the programming of movements is reconstituted through the creation of an artificial and rudimentary nervous system.
>
> (Leroi-Gourhan 2003: 243)

The rise of the machine then begins with the liberation of fingers and continues into the civilisation of the automata. As the prehistoric world emerges into the historic the 'motor function' and technique continue to develop in 'machines such as the bow or the crossbow, in snares, pulleys, millstones, cranes, and transmission cables' (2003: 245). Each of these was not an adaptation to nature but the adaptation of nature submitted to the gratification of our hand:

> If the hand of the earliest anthropoid had become a tool by adaptation, the result would have been a group of mammals particularly well equipped to perform a restricted series of actions: It would not have been the human being. Our significant genetic trait is precisely physical (and mental) *non*adaptation: a tortoise when we retire beneath a roof, a crab when we hold out a pair of pliers, a horse when we bestride a mount. We are again and again available for new forms of action, our memory transferred to books, our strength multiplied in the ox, our fist improved in the hammer.
>
> (Leroi-Gourhan 2003: 246).

Leroi-Gourhan argues that these machines have their origin in the biological organism of humans and are in fact organic machines, albeit a mutation of the human organism and a substitution of the organ of the hand by a technical device (2003: 246). But these devices are also substitutes for the 'animal machine' of subjected biological species like horses. Much like the technical development of the medieval period as a reaction to the freeing of slave labour as Marc Bloch noted, the liberation of animal power (which becomes a fuel or food rather than a labouring device) is the motive-power of the machine age. For Leroi-Gourhan:

> The *animal machine* requires a good deal of muscular participation. Motor function is "deflected" to drive the animal motor, but it remains considerable. Moreover the efficiency of the animal-driven machine became stabilized very early on and at a rather low level: The number of horses does not increase the speed of the vehicle nor, within certain limits,

> their resistance to fatigue. The relationship between humans and their exteriorized force is altogether different in the *automotive machine*, including even the simplest water-driven piledriver or mill. Having set the process in motion, the hand no longer intervenes except to feed or to stop the machine. The operator can increase the machine's power or distribute it among machine tools which will perform all the operations for which human intelligence has designed them.
>
> (Leroi-Gourhan 2003: 246)

Ultimately, the automotive machine will subordinate its own creators to itself as appendages – 'the robot gravitating like a constellation around the human of flesh and blood' (Leroi-Gourhan 2003: 248). He sees the development of mechanics itself, specifically clock-making, as a form of technological evolution comparable to animal evolution – specifically in the development of more complex speciation (2003: 249). The distancing of humans from nature and the 'exteriorization of the brain' (2003: 252) was about the kinds of materials from which the new prosthetic forms were fabricated, and also the type of fuels that were used to power the mechanics of the new machines. For Leroi-Gourhan the fundamental and most important human development was the invention of the iron-foundry and thereby the use of coal and steam to accelerate the industrial anthropocentric domination of nature. This was largely a question of force and power and was intimately related to the birth of the industrial proletariat:

> Railways and the emergence of a proletarian working class, to name only two of the immediate consequences of the freeing of driving power, have had a direct effect on the entire organization of our species. The adjustment of human individuals, whose brain and physical frame are still those of CroMagnon man, to the new conditions has involved an ever-increasing degree of distortion.
>
> (Leroi-Gourhan 2003: 252).

This we can see in the distortions of human musculature as they work around the machines but also the self-distancing and alienation, in a Marxist sense, from what might have been the 'unity' rather than the 'schism' of the human interior. This again returns Leroi-Gourhan to the technical, 'third' memory of the machines:

> Like tools, human memory is a product of exteriorization, and it is stored within the ethnic group. This is what distinguishes it from animal memory, of which we know little except that it is stored within the species. Animal, human, and mechanical memory differ from each other in some essential respects. Animal memory is formed through experience within narrow genetic channels prespecialized by the species, human memory is constituted through experience based on language, and mechanical

> memory is constituted through experience within the channel of a preexisting program and of a code based on human language and fed into the machine by a human being. Mechanical memory is not without some points of resemblance to animal memory: A kind of species-related preconditioning exists in all types of machines, but the operating program is dictated in a wholly instinctive manner because it materially preexists the action, whose every twist and turn is plotted in advance.
>
> (Leroi-Gourhan 2003: 258)

The process of anthropogenesis, the *self-making* and bringing into being of their own humanness, then is entwined with their human *making*. This exteriorisation having its origins in their interior being is also part of re-making that interior being through the consumption of fuel and products and concepts that history and nature have distributed to them, or have been grasped by humans themselves. But it also signals the end of the biological being of the human. This is what we might term the phase of post-humanity, what comes after *Homo sapiens* has totally distanced from and altered their animality and of nature more generally. But it is also the case that there is another concept of the post-human, one which displaces humans not by machines but topples machine civilisation with humanity itself. The extinction-level events of ecological catastrophe and disaster might end that Anthropocene epoch just as much as the machines submerge humanity. Memory itself might be entirely transposed into machines and distanced from its biological origin. But for Leroi-Gourhan:

> We must therefore expect a completely transposed *Homo sapiens* to come into existence, and what we are witnessing today may well be the last free interchanges between humans and the natural world. Freed from tools, gestures, muscles, from programming actions, from memory, freed from imagination by the perfection of the broadcasting media, freed from the animal world, the plant world, from cold, from microbes, from the unknown world of mountains and seas, zoological *Homo sapiens* is probably nearing the end of his career. Physically the zoological species has a certain future…Species do not age, they are transformed or disappear. In any case the future that lies before us goes far beyond the rate of human sociotechnical evolution.
>
> (Leroi-Gourhan 2003: 407)

The eclipse of human beings at the hands (digits) of the machine is a popular motif within science fiction and in new Cyborg studies. The compelling vision of the human-machine ensemble generally entails either a utopian or dystopian romantic vision of humans against machines. Vaccari and Barnet, in their analysis of Stiegler and by extension, of Leroi-Gourhan, have posited the idea of a retrospective 'robot-history' documenting their own origins:

> Tracing that vast history at the speed of light, the historians might even come to the conclusion that there is no natural class that can be called 'technology' or 'machine'. Like Darwin, they will believe that a species is a historical artefact, and that phyla can only be reconstructed historically, that is, according to certain family resemblances that can only be retroactively ascertained. Phyla are historical artefacts, yet unlike biological phyla, machinic phyla respect no natural kinds, no borders between human faculties and machine code. The robot historians will incorporate transgenic seeds, test-tube babies, strawberries, germs, fishes and forests into the Great Family of The Machine. Machine history will not be natural history of humans and their wars and famines and love stories, but a pan-machinic, universal history of what Stiegler calls "organised, inorganic matter".
>
> (Vaccari and Barnet 2009: 3)

The subsumption of the human and the organic matter into universal machine history rather than vice versa signifies to which species might be ultimately triumphant. Vaccari and Barnet also point to the relationship between the work of Stiegler and that of Derrida partly suggested by Stiegler's intense engagement with Derrida's work in the first volume of *Technics and Time* (1998). In all ways this rests on the original idea of the technical memory that has its origins with Leroi-Gourhan,

> …the possibility that human memory is a stage in the history of a vast machinic becoming. In other words, these future machines will approach human memory (and by extension culture) as a supplement to technical beings. What matters is the transversal movement that posits human memory as a process of recording that exceeds the boundaries of the mind and the logic of the human. Indeed, a process that is distributed across time and space, which explains the wonderful plasticity, resilience and adaptability of technical codes as opposed to individual memory; or even genetic memory, which is often fragilely bound to the fate of environments. The future machines will divine here a posthuman archewriting that is as primeval as life and founds the logic of inheritance, memory and transmission.
>
> (Vaccari and Barnet 2009: 10)

The relation then between the hand and the loom is one that is about both connection and separation and about defining the bounds of being human – which is extended into this 'machinic becoming'. Perhaps the fascination lies in the fact that technics is implicated into what we know about anthropogenesis and the birth of the human and also in its end, completion or extinction. This has a number of consequences for social theory. First, that the anthropogenic observation point is being displaced. The idea that humans are at the pinnacle of the zoological taxonomy is one aspect of this. This is the retreat back into the idea of human animality of Critical Animal Studies and the reassertion of the rights of other sentient species and ecologies. It is therefore a form of post-humanism. On the other hand, the idea of the subsumption of human identity and memory into the 'machines' moves the human further away from its animality and almost into a pure technics. It is also thereby a form of post-humanism. Each of these political and theoretical positions offer certain types of recipes, prescriptions and programmes. Second, the problem with the structural anthropology of thinkers like Leroi-Gourhan lies in the questions of attribution of meaning to huge classes of data. Ideas of order, succession, grouping are central to Leroi-Gourhan's thesis about aesthetics and prosthetics but in these ideas have been undermined in recent years by the reassertion of a comparative, ethnographic anthropology which is less about ordered structures and more about understanding prehistoric societies by analogy with contemporary or near-contemporary shamanic, tribalised societies (Lewis-Williams 2002, Mithen 1998). Third, there is a pejorative aspect to both animality and technicity which shapes the observation of each of these concepts. Stiegler is particularly well-understood as an 'industrial' thinker (Stiegler 2014) while Castoriadis is fundamentally critical of the human use of technics and machines, seeing them as opposing human autonomy and self-making. The different readings of Leroi-Gourhan rest upon the kind of society that one wants to prescribe as the solution to the 'catastrophe' of the machine.

The locomotive as dream

John Berger once asked a question about the centrality of the railways in human imagination: 'Before the railways were built, what took the place of stations in people's dreams?' (1992: 92). Railways have an overwhelming cultural power and significance in modernity. From the locomotive and fabricative hand the human collective has developed an immense number of machines. As it has done so it has made, transformed and potentially ended itself as a human form. In the next two sections of this chapter we begin to discern the movement of the locomotive itself, the steam-engine, diesel car, cattle-wagon, troop train. The importance of the train lies in that sense of separation testified to by Leroi-Gourhan. The train separates humanity out from itself, from its bipedal relationship to the earth, and its products from their original locations. It empowers the individual human hand and foot many hundreds of times and propels them into and through locations far distant from their spaces of birth. The train is a machine of dislocation and extraterritoriality even as it has an intimate relationship to the geology of the territories it traverses. Hundreds of thousands of the labouring dead are buried in its wake. It is also an engine of disappearance. Millions of people were transported in its cattle-trucks to specific sites of extinction: Auschwitz-Birkenau, the battlefields of the Somme, the Stalinist camps of the Vorkuta and Kolyma. But of course, it is also a 'romantic machine', signifying separation, reunion and love. It is an engine of movement, of connection and of accumulation. It is a technology of mobility; of animals into slaughterhouses, armies into war, 'sub-humans' into camps, lovers into one another's arms. At its birth is a rejection of grasping for fabrication, of the world and oneself. For Bernard Stiegler:

> Originary man is practically immobile: finding nourishment under an oak and "finding his bed at the foot of the tree which afforded him a repast", he does not change places, nor move across space, let alone desire to possess the whole of it or even a constantly greater part of it. He is not moved excessively because he is not truly moved, he has no passions: he has natural, vital needs, which he can satisfy in reaping the fruits of nature close at (his) hand—this hand that only grasps without manipulating anything, an organ of grasping rather than of fabrication.
>
> (Stiegler 1998: 114)

The propulsion of the train, its motive-power has its origin in desire and possession. A product, as Stiegler says, of passion rather than human need.

The railway had its origins in the medieval period in mining technologies. Medieval railways were often made out of wood and were driven by human labour or animal power (Lewis 1970: 8–9). They were already organic machines: their raw materials came from trees and their motive-power from animals. Their extractive function multiplied with the development of the steam engine and its use in rail and mining technology by Stephenson, in the ensuing years it become a mobile engine of transport for humans. But what was the essence of these technical advances? Rodney Dale in his history of the early railway notes that: 'A "locomotive" is a machine which moves under its own power (or, as we now say, under its own steam). Steam was the power of the Industrial Revolution, as coal was the fuel – both literally and metaphorically', (1994: 15). Railway lines emerged as veins of the industrial system. They were the incarnation of mobile capital, engines for accumulation and transportation of labour power and eventually of the leisure classes of European populations. The engineering of the infrastructure was immense: 'Barren and blasted heaths, inaccessible mountains, impassable estuaries and places totally devoid of inhabitants – all had a line', (Dale 1994: 51). Railway 'mania' and speculation fuelled its industrial production and rail catastrophes became central fictional motifs of Victorian and Edwardian literature. But the emergence of the rail 'system' was essentially a product of accident and contingency. There was absolutely no English national plan or policy of rail extension for example (Bailey 1995: 10). The lines connected only by accident in many cases. There were multiple rail companies often connected to iron foundries and coal mines each with their own systemic infrastructure and with little intention of creating a national or international system beyond their own accumulative desires.

Ultimately the railways would literally span the world enmeshed in profound experiments in engineering and architecture as well as the development of the trains themselves. Keith Lovegrove's work on the design of rail technology has noted the immensity of the train project: 'The locomotive has shaped and parted nations, tamed wildernesses and built empires. Populations of vast metropoli and isolated communities have socialized and cross-pollinated with the help of the two rails that never meet', (2004: 7). Yet it is in the aesthetics of the locomotive object itself that its power lay to shape the social imaginaries, desires and fictions of its period. For Lovegrove:

> During the 'Golden Age' of steam, engineering dictated form and, by extension, aesthetics. The defining visual characteristics of the steam locomotive are a collective accident of design...Decorative elements were limited to a filigreed crown on the smoke stack or a polished brass dome on the firebox; the beauty of Victorian engineering was its exoskeletal and random styling.
>
> (Lovegrove 2004: 20)

The bricolage and assemblage of the locomotive object was itself a product of the social imaginaries of its day and it profoundly shaped the aesthetics of the social world in which it was enmeshed. The function of this exoskeletal machine was to transport, accumulate and power but it also performed an aesthetic function and act for the witnesses and observers of its passing, through cities and fields and in dreams where the train appeared as an engine of desire or fear. Lovegrove argues that the locomotive was constantly metamorphosing by the entanglement of industrial design with engineering possibility:

> By the mid-nineteenth century, production of steam locomotives reached the hundreds, and advances in engineering brought about a metamorphosis in the appearance of engines. The inclined and vertical cylinders that had been a distinctive feature of early experimental models were now mounted horizontally and hidden from view next to the multi-tubular boiler.
>
> (Lovegrove 2004: 20–25)

The experimental laboratory of the locomotive and the rail infrastructure was equally a laboratory amid social and natural landscapes. Even though the locomotive infrastructure of engineering and architecture was limited by nature in its geology and material capacity it also transformed natural habitats, created cities where none had been and drove tunnels through mountains. It transported humans who had once been rooted for their entire lives in specific local landscapes across the globe creating new modes of ethnogenesis and globalised cultures.

The 'romantic machine' gradually dispensed with its organic elements and like all other machines subordinated its humans to its own machinic will, the sentient cogs servicing the system as a whole. But the great railway stations and track systems became part of the global mythology of human possession and dispossession, of disappearance and reappearance and of individual human love and desire. John Berger, in his study of love, has noted this aspect of the locomotive culture:

> Of all nineteenth-century buildings, the mainline railway station was the one in which the ancient sense of destiny was most fully re-inserted. Stock exchanges, banks, hotels, theaters, courts were built as pretenses, or, to put it another way, they were already dreams. The railway station – whatever the extravagances of its "decorative" architecture – remained stark. And it remained so because it was a site of arrival and departure, where there was nothing to muffle the significance of those two events. Coming and going. Meeting and parting. Dreams welcomed the railway station so readily since it was already – in other forms – a familiar. The Greek word for "porter" is *metaphor*. And this is a reminder of how deeply the act of transporting, of despatch and delivery, is intrinsic to the imagination.
>
> (Berger 1992: 92)

Junctions, conjunctions, potentialities, arrivals, departures. The use of the fictional motif of the 'station' in film and literature was perhaps related to something which the system itself intimated – the mode of abstraction between the single individual passing through the vast halls and transported across huge distances and the magnitude of the system itself that made this possible. Berger expressed this entanglement and contact between the individual and the abstract:

> In a railway station the impersonal and the intimate co-exist. Destinies are played out. The trains run regularly, according to printed timetables. The lines are inexorable. But for each passenger or for each person who comes to meet or see off a traveler, the train in question has its own portent. The portents can be read close-up, in faces, in details of luggage, in the welcomes and partings as people embrace on the platform.
>
> (Berger 1992: 93)

The social imaginary of the vast machinic structure, of the monstrous locomotive engine, of the 'inexorability' of the system as opposed to its individual humans makes the power of this technological system manifest.

Simon Bradley in his history of the British rail system has outlined the cultural power of the railway:

> The lines themselves – the "permanent way", in railway terminology – carried a historical charge of their own. Landscapes that had barely altered since Shakespeare's time were suddenly scarred by gigantic embankments, or punctured by tunnels so long that the very survival of the enginemen amid the smoke and fumes were sometimes imperilled. Tens of thousands

> of bridges and viaducts carried the new routes across roads, rivers and streams, flood plains and estuaries.
>
> (Bradley 2015: 5)

The veins of this new system utterly transformed both the natural and the cultural landscape of the British nation even as it had no single governing plan, aspiration or programme (2015: 9). The metaphors and vocabularies of the steam-train world became, as Bradley notes, the metaphors and vocabularies of our whole social world (2015: 15). The single time standards of the network developed and were eventually solidified in a new unitary system that eventually joined up all the old pioneer lines into the British rail system. Although haphazard in development, it created a unified, automated, machine-system out of its contingent and accidental historical origins. Each station had their 'Clock' (itself becoming a stock metaphor of romantic film and novel) which was subordinated to Greenwich Mean Time at the same time as it subordinated the railway labourers as appendages to it and compelled the individual traveller into subjection. The development of this new system of industrial timing would transform our conceptions of arrival and departure and indeed change our view of the nature of time itself. Time became not something observable in the seasons or in the patterns of nature but part of the vast machinic network of Capital and its clocks (see Postone 1993 for the fullest elaboration of time and social domination). For Bradley:

> To show the time in this way implied a claim to the kind of authority that had traditionally resided in municipal and communal buildings, guildhall and town hall and church steeple; it was a significant embodiment of the power of capital, and a reminder of the extent to which conditions and standards were to be imposed forthwith from outside the community.
>
> (Bradley 2015: 21)

Bradley argues that the strict timetables and regulations of the system also transformed the ways in which time could be subdivided into its elements before which the archaic manners and customs of viewing time became 'doomed and impotent' (2015: 21). The emergence of this new, divided timetable and regulated universe signals the end of the epoch of the 'prerailroad world' (Bradley 2015: 21).

Like any new civilisation the railroad system was, however, a mixture of different modes of production and inheritances from the past. What seemed profoundly modernistic in the Victorian period such as the use of multiple oil lamps and then gas (Bradley 2015: 33–34) seems to us perhaps as antiquarian. As Bradley says:

> The glass in these early windows was not very like the coated, toughened and tinted safety glass of today's carriages. It had the same ripples and watery imperfections that were found in the window glass used in the

> ordinary houses of the time…So it was through slightly distorting apertures that the first railway travellers viewed the world as it accelerated and decelerated strangely past them.
>
> (Bradley 2015: 28)

This acceleration and deceleration, the shift in our cultural view of time, movement and landscape would become part of both the comic and the tragic vision of the 'novel' itself fully emerging at the same time as the locomotive culture and seen even then as enmeshed with its development.

But the train also was about 'The classes in motion' (Bradley 2015: 54). The structure of the trains into various classes and estates, the use of private trains by royalty and capitalists and even the structure of waiting rooms made visible the 'order' and structure of the Victorian period. Not only did the locomotive move labour around, it also subjected it to a regulatory time system and made material in the very being of the interior of each worker a subjection to the capitalist order. By ranking its individual humans the locomotive system ensured the visible reproduction of its own physical and signifying order into the human mass. Indeed, it made the strata of the capitalist system seem preordained and even part of the 'natural' order of things.

The locomotive as nightmare

The locomotive appeared in dreams and romantic novels as an engine of desire but it was equally apparent in the individual and social nightmares of the nineteenth and twentieth century. The train was enmeshed in the literary devices of disaster and catastrophe (Fyfe 2015, Matus 2009). By the time of the industrialised annihilation of the First World War the train was being used not just for transportation but for disappearance. The locomotive was central to German plans in the Franco-Prussian war and to the subjection and almost total elimination of the indigenous peoples of the Americas. On the domestic front, in Britain the rail system assisted war overseas (Gittins 2010), as well as being central to the automotive and percussive power of the army on the Western Front (Aves 2009) and in alliance with the British Expeditionary Force in 1939–40 (Aves 2012). But the railway was simply the machinic extension of technical civilisation into the trenches. It extends the subjection of the human appendage to its will as we have seen but in the case of war it also eliminates its human appendages as collateral damage while actively trying to destroy both the sentient appendages of the opposing 'forces' and their machines of 'armed desire'.

Daniel Pick, in his analysis of total war, has tried to elaborate the relationship between the rise of machines and the prosecution of war:

> The figure of war is torn between such different discursive possibilities: it is seen to provide coherence, boundaries, meaning but also to erode the identity of the structures and forces that inaugurate it. War has been

> conceived as cohesive, purposive, stabilising, but elsewhere as the consequence of a mad and uncontrollable machine or machine age; as protean and slippery, amenable to many definitions over time, fragmented by its countless national and local forms.
>
> (Pick 1996: 7)

What the trains allow is the collection and accumulation of vast numbers of individual labourers and their transportation to a highly specific landscape – the war 'field'. As Edith Wyschogrod argued in her analysis of Hegel, Heidegger and mass 'death-events' the mass death is a complete distortion of nature. She argued that 'What is unprecedented in the new phenomenon is that the means of annihilation are the result of systematic rational calculation and scale is reckoned in terms of the compression of time in which destruction is delivered', (1990: x). The timetabled and regulated system of the railroad is the precursor to the delivery of the mass death-event. It is a compression of time but it is also the compression of space to the highly specific locations of assault and destruction whereby this vast abstract system delivers its flesh to the place to be then torn apart by what Ernst Jünger called the 'storm of steel' (2004).

Jünger, a friend of Heidegger, and obsessed by both insects and machines, would also elaborate in the period before the Second World War a fictional vision of slaughter and the death camps in his conservative opposition to Nazism (Jünger 1970). The vision prefigured actual mass deportations and mass death-event of the Holocaust initiated by the Wannsee Conference in 1942 in which the train and its trail of cattle-trucks was so decisive. Gerlach in his analysis of the logics of deportation documents the transportations (2000: 112–113). He argues that there was a distinction between what he calls the 'utopian' and the 'practical' potential programmes for deportation. For Gerlach: 'What was decisive for the actual realization of mass murder plans were the intermediate steps between the utopian emigration and extermination schemes, on the one hand, and liquidation programs that could be practically implemented, on the other', (2000: 139). Similarly, Anne Applebaum has examined the logistics of the Stalinist death camps with the transits being the 'most inexplicable aspect of life in the Gulag' (2003: 160). Often, and ironically, the trains to the camps were disguised as food trains carrying bread or vegetables (2003: 160). Martin Gilbert (1986: 306), in his authoritative history of the logics and logistics of the Holocaust, documents the case of young Luba Krugman who watched the transit trains passing through Chorzbany, near Lublin:

> At night the rumbling of the train woke me up. I rushed to the door and found it wide open, with father watching the dark monster, which had only one brightly illuminated car window behind the locomotive.
>
> "It's a real Cyclops", I said.
>
> "I can't hear you", replied father.

> Our voices were drowned out by the clatter of the wheels and the penetrating whistle of the engine. I counted sixty-two cars.
>
> (Gilbert 1986: 306)

In effect, what Luba Krugman witnesses is not the train as an engine of production but of *de*-production. The Final Solution, initiated by Wannsee, accumulates bodies in order to concentrate and compress them in a specific space and then destroy their physical being (in doing so using the remnants as products for the German war effort). The disarticulation of the physical being of the Jewish people means that re-articulating that experience in memory and in documentation is a project of utmost import against revisionism and denialism.

Perhaps the most important analysis of the train as mode of transportation and destruction lies in the work of Simone Gigliotti (2006, 2009). For Gigliotti, the experience of deportation during the Holocaust has to be understood as a form of traumatic 'spatial, psychological, and sensory assault' (2006: 256). At the centre of this transit and this encounter with trauma is not the locomotive engine but its secondary appendages – the cattle cars. Gigliotti asks:

> How has the cattle car experience been constructed as a traumatic trope with hermetic resonances in innumerable testimonies after the event? Testimonies reveal that the journey itself did the work of stripping— emotionally, physically, and spiritually— those markers commonly perceived as boundaries between civilized and subhuman existence: spatial distance between people, the separation of private and public activities, and the ability to act on informed ethical choices.
>
> (Gigliotti 2006: 262)

We have here that sense of separating and distancing that we have observed above. In this case it separates out locations and distances the human from the sub-human as well as distances around politics and ethics. Gigliotti's attempt to reground the experience of the transit is a peculiarly sensual one and about the intimacies of the body: not just what is perpetrated upon that body but 'olfactory' witnessing. This entails the smell of spaces, of cars and of camps. It is a witness by the body itself, of itself and of other bodies and places (2006: 265–266). The 'deportation travel' of the cattle car is metaphorical as well as literal: 'The metaphor is migratory: it has passed through decades and generations, and across ethnic groups and wars, yet it is also immobile, a referent of historical specificity and entrapment', (2006: 271). The specificity of the experience is contrasted with the wider metaphorical and literal experiences of spaces of exception, disempowerment and destruction (see also Lozowick 1991).

Gigliotti extends and develops this further in her book on the train experience. She argues for a 'sensory witnessing' as a central aspect of survivor memory (Gigliotti 2009: 3). It is almost impossible to document the hidden genocide locked within the cattle car through the usual historical methodologies so the intimate history of the body becomes the most important way of understanding the experience of entrapment. The regulatory and timetabled aspects of the Wannsee protocol are of course present in official documents but the experience of the embodiment of captivity is something to be found elsewhere. For Gigliotti:

> They are the victims' history of Nazi deportation policy, which represented the immobilization of personal mobility…I read deportation's trauma as a sensory and embodied history of train experiences that radicalizes nineteenth-century responses to train transit. These responses were grounded in spatial and somatic trauma. They included changes to perception, distancing from the natural world, and sensorial disconnection from landscapes because of mechanized transit.
>
> (Gigliotti 2009: 2)

The disconnection between the basic human metabolisms of captor and captive initiate an awareness of a counter-archive of sensory embodiment which testify to the history of victims rather than perpetrators (2009: 3). Like the dark hold of the world of the slave ship (Hudson 2016b), the cattle car offers us a reassertion that these mobile camps of imprisonment were full of humans with senses and with bodies. The embodied corporeality of the captives in the 'Holocaust transit' (Gigliotti 2009: 19), congeal a different set of truths to those of the guards (2009: 21). Fundamentally Gigliotti's work displaces the location of the camp as 'symbolic geography' in favour of the embodied experience of mobility. The experience was more about movement than it was about destination (2009: 23). The relationship of the body to the machine in movement then returns us to the hybrid metabolism, one that is in rift, drift and separation.

Metabolic vehicles

Paul Virilio has pointed to the metaphor of the 'trainwreck' for our accelerated modernities (Redhead 2004: 46). And as Steve Redhead has noted: speed is intimately related to military acceleration in Virilio's social theory (2004: 2). Part of Virilio's work is to understand the nature of what he calls 'metabolic vehicles' (2006: 96). To understand these vehicles, it is necessary to address his theory of movement and its relation to modernity. For Virilio, the beginning lies in the fictions of the whale ship and the Cetacean holocaust that we have already examined:

> *Gordon Pym* and *Moby-Dick* are only the anticipated narratives of the nuclear cruise. The strategic submarine has no need to go anywhere in particular; it is content, while controlling the sea, to remain invisible. But its hourly fate is already sealed. Furthermore, as soon as the fleet in being

> becomes a fundamental given of the Right to the Sea, the explorers, discoverers and raiders of every stripe, while continuing to seek uncharted lands, equally adhere to the invention of passages, in other words to the realization of the absolute, uninterrupted, circular voyage, since it involves neither departure nor arrival. The loop of no-return is traced in advance by the circular or triangular nautical routes of European mercantilism.
>
> (Virilio 2006: 65)

The passages of modernity are essentially ones of drift, without destination. The submarine is mobile around the ocean, as migratory as the whale, as circular as species in movement: like the 'veins' of the whale passage or the middle passage of slavery (Hudson 2016b). These circular movements are essentially militaristic migrations although the latter are often across material territory rather than through ocean. In 1917 the war in France was about a series of voyages, but in cattle cars rather than ships. As Virilio argues: 'The voyage of the masses once more took them from the street to the railroad… After that, the military livestock quickly jumped on the cattle wagon, but everything was over very fast', (2006: 77). They were all cut down by the storm of steel, the 'rapid engines' of war circulating territories like ships plying their way across the sea (2006: 78). The metabolic vehicles of humans could be 'boarded' like ships (2006: 96–97). But these metabolic vehicles of labour were hybrids of the organic and inorganic. The 'animal proletariat' were used as labour for war and accumulation itself driven by the 'vehicle-bodies of horses' (2006: 106). But the human metabolic vehicles are bodies controlled by 'disquieting tenants' controlled by alien intelligences of power and war (2006: 108–109). The metabolic vehicle is only truly possible when there is a specific entanglement of the organism with machinic structures. As Virilio says:

> For the dromomaniac the engine is also a prosthesis of survival. It is remarkable that the first automobiles, Joseph Cugnot's military trolley of 1771, for example, were steam-powered, already situating themselves at the limit of the animal body's metempsychosis, relay of historical evolution: the limit of the passage from the metabolic vehicle to the technological vehicle, spilling its smoke like a last breath, a final symbolic manifestation of the motor-power of living bodies.
>
> (Virilio 2006: 114)

Virilio points to the dispersal or the subsumption of the metabolic aspects into the purely technical, just one passage in the robot history as Vaccari and Barnet indicated above.

Daniel Pick has argued that there is a contiguousness between the railway as transports for soldiers and the railway as a mobile death camp for animals:

> Take the modern assembly-line slaughterhouse. Just as rails now transported soldiers to battle, they bore animals to their death. The coming of

> the railway in the first half of the nineteenth century had opened up new possibilities in the assembly process, enabling the reduction of delays and friction between each productive stage.
>
> (Pick 1996: 178)

Blaszak et al. have pointed to the centrality of slaughterhouse for the development of the Chicago railroad network (2014). Dominic Pacyga has documented the opening of the Union Stock Yard, Chicago, at Christmas 1865 (2008: 153). He notes that the 'great bovine city' of Chicago was decisive for the whole of the railroads of North America: 'Within four years of the Christmas Day opening of the stockyards, railroads spreading out from Chicago would connect both coasts and create an enormous hinterland for Chicago's meat industry, further enhancing the city's role as the Emporium of the West', (2008: 153–154).

Paula Young Lee, in her critical analysis of slaughterhouse modernities, quotes Upton Sinclair's vision of the stock yards: 'Now and then a visitor wept, to be sure; but this slaughtering machine ran on, visitors or no visitors. It was like some horrible crime committed in a dungeon, all unseen and unheeded, buried out of sight and out of memory', (2008: 1). As Lee notes, Sinclair's novel laid bare the labour relationships of the yards, the use of tainted meat and intimated that consumers might be eating the meat of human flesh of workers who had been trapped in the slaughterhouse machinery (2008: 1). In her analysis she argues that the slaughterhouse was 'one of the exemplary institutions of the nineteenth century…' (2008: 2). It was an impersonal machine where human conscience and ethics were evaded as the bodies of animals brought in on the trains and cattle cars were de-produced, disarticulated, disassembled, dismantled and turned simply into product. As Lee says: 'The slaughterhouse did not assemble a new machine along a rapidly moving line but disassembled a pre-existing animal, part by bloody part, until it re-emerged as a carcass without life, skin, or entrails', (2008: 4). It was literally a machine for death and extinction produced by the vast appetites of the cities. She references Corbusier's architecture of utility: 'the house-machine and the slaughter-machine share an umbilical connection to the same point of origin: the modern metropolis', (2008: 4). But like Sinclair, watching the machinic slaughter of organic beings and seeing ethics in suspension it is the consuming being that revels in the slaughter rather than rationality. As Lee argues: 'For just as it is the machine that kills, it is also the machine that watches: the self-interested spectator of these events is the stomach, not the conscience', (2008: 6). The emergence of the desiring consciousness with the hand as its mode of extraction and accumulation does not tear the meat. It has developed a whole prosthetic civilisation to tear and consume the meat from the beings on the engines of mobility.

6 Memory, animals and nature

In this chapter we reassert the power of social memory specifically with regard to the history of nature: a history generally composed, written and categorised by its human contestants. There is no 'equal to' between the Cetacean and the Jewish holocaust, neither does the 'other' holocaust of the whales mitigate the unique and radical horror of the human holocaust. The latter was not just 'one amongst many' genocides but something utterly propelled by the fear and loathing of a specific people that had developed a set of motifs and amalgams over centuries. The importance of the Cetacean and the animal holocaust more generally lies in its prefiguration of the human one: the use of highly rationalised and weaponised methods of slaughter and transportation, the visual imagery and tableau of utter destruction, the pieces of body cast into the fire and the dispersal of the memorial remnants of the massacres. These were both holocausts that consumed the bodies of the fallen, for utility and for the gratification of destruction. Yet, there are also significant differences. Although the bear and the whale were tormented for utility and for pleasure, their wildness and savagery were seen as something precious and unique. The hunters respected their prey. The Cetacean holocaust and the 'Animal Treblinka' did not take place in order that one part of nature be utterly destroyed because of its 'unnature' and its monstrousness, even less so because of the fear that it might contaminate and breed with humans. The animal holocaust took place because of accrual, accumulation, ownership, capital. Utility and industry were an offshoot of the human holocaust, a by-product without being its immediate rationale. The animal holocaust was simply one of extraction and mercantilism. The captivity and destruction of animals was not born of hatred but of curiosity and accumulation.

Like Scoresby's polar bear, the animal in captivity becomes something which is stared at and loved in its cage as well as something tormented, observed and classified. John Berger, in his remarkable vision of the Bear, dreams of himself and the Bear together as antagonists in a room. The Bear was chained to the wall. They conversed, they wrestled, they danced. Then Berger realises that the Bear is not chained to the floor. But the Bear does not know that he is unchained (1993: 135). Yet: 'In every liberation, whatever happens later, there is a beauty. This beauty has a kind of abandon to it. Bear was dancing with more and more abandon. I had been mistaken. Bear knew. Perhaps had known from the beginning', (1993: 135). In Berger's story the Bear forgives (1993: 137), but we cannot know whether the Bear has forgiven because its social memory is encased in the cultures of its captors and contestants. In what sense can we create a fragmentary social memory rooted in the embodied experience of the destroyed animals?

Documenting the Arctic and Antarctic

The social memory of whaling is an archive over and against the social memories of the whales and other oppressed species. Scoresby's observations of and documentations of the Arctic were enmeshed in literal machineries of destruction. These machines were validated by observation, surveillance and collection: an entire ideology of animal subjection and destruction. At once, taxonomic classification is a way of uncovering, and in literal terms 'flensing' the whale in order to understand its interior being – all the better to understand the dominion of God and human over animals and to subject further animals to this process. This dominion was not wielded simply for its own sake but because it was enmeshed in utility and accumulation of the new empires of capital (see Lucas and Edwald 2015: 231). The congealed capital inherent to the whale was central to the mercantile systems that were reaching out into the North and South Atlantic worlds; seeing, listening and accumulating/collecting.

Edmund Wilson, on what would be later understood as Scott's 'first expedition' kept a series of diaries that were later collated and published. As he travels across the South Atlantic world and into the great ice barrier of the Antarctic, Wilson recounts the tracking and the sighting of the whales (1966: 40). He has a series of sightings. Of Finback whales beyond Cape Town (1966: 64) and perhaps more dramatically, over a hundred Grampus whales together along the ice barrier (1966: 108). What is remarkable about the diaries is that Wilson is entering an 'alien' world where his whole conception of nature and the universe is transformed. Further, his diary, collected upon an 'official' journey, is itself but an official and unofficial archive of the contacts and encounters between human beings and other species including the whale, the seal and the albatross. This archive of encounter then creates a possibility of being 'archived against' – of new counter-archives observing humans from the position of the whale and the seal. But is this possible from the perspective of human beings and indeed, how can we replace human perspectivalism being human ourselves?

Some beginning might be possible by re-encountering another journey across the South Atlantic: the narrative of Arthur Gordon Pym. Poe's great fictional work is enigmatic but Pym, like Wilson, encounters albatrosses and seals in the great Southern Ocean (Poe 1998: 112–114). Indeed, the name of the ship is itself the *Grampus* (there was also a *Grampus* in the great naval mutinies of 1797). Locked inside the belly of the whaleship, Pym's encounter with the whales is framed as a journey of human destruction within the interior of the creature itself. Pym's history of navigation in the southern latitudes (1998: 120–125) leads to a bitter moment of human destruction. He documents the laceration and the killing of a kind of Arctic bear which they then eat upon the ice (1998: 127–128). This moment of disruption and destruction, of empire-building, consumption and sating of appetite in a lonely place that had never before encountered humans is a moment of horror. The contact and encounter of two modes of being end in the destruction of one of the species. This is a singular moment of one bear's destruction but this could ultimately come to represent the dissolution of the whole species in the ecological catastrophe of modernity. The destructive humanness that the tale exemplifies is not just that of the destruction of animals but of humans wantonly killing each other, revelling in their own destruction and destructiveness as a species. Yet the ending of the tale is profoundly mysterious. Pym and the others meet their fate upon this strange continent:

> And now we rushed into the embraces of the cataract, where a chasm threw itself open to receive us. But there rose in our pathways a shrouded human figure, very far larger in its proportions than any dweller among men. And the hue of the skin of the figure was of the perfect whiteness of the snow.
>
> (Poe 1998: 175)

Whether angel or ice formation, or even the Brocken Spectre over the abyss, this semi-human figure portends the destruction of human beings themselves because of their own hubris.

It is with the idea of fiction, fancy and fairy tale that perhaps the greatest account of Arctic history begins: that of Fridtjof Nansen. Nansen's magisterial work documents the whole extant history of Arctic exploration beginning with the classical world. For Nansen, the exercise of exploration is centrally about the accumulation and utilisation of forms of knowledge:

> By tracing how ideas of the Northern World, appearing first in a dim twilight, change from age to age, how the old myths and creations of the imagination are constantly recurring, sometimes in new shapes, and how new ones are added to them, we have a curious insight into the working of the human mind in its endeavour to subject to itself the world and the universe.
>
> (Nansen 1911a: vii)

The 'subjection' of the world is at once an exercise of observation and collation and then ultimately of physical subjection, albeit in the harshest environment on earth. For Nansen:

> In the beginning the world appeared to mankind like a fairy tale; everything that lay beyond the circle of familiar experience was a shifting

> cloudland of the fancy, a playground for all the fabled beings of mythology; but in the farthest distance, towards the west and north, was the region of darkness and mists, where sea, land and sky were merged into a congealed mass – and at the end of all gaped the immeasurable mouth of the abyss, the awful void of space…With unspeakable labour the eye of man has forced its way gradually towards the north, over mountains and forests, and tundra, onward through the mists along the vacant shores of the polar sea – the vast stillness, where so much struggle and suffering, so many bitter failures, so many proud victories, have vanished without a trace, muffled beneath the mantle of snow.
>
> (Nansen 1911a: 1)

The historical procession of 'struggling, frost-covered figures' (1911a: 1–2) traversing the icy, polar darkness is about the 'eye' – a recurring theme in Nansen's work. The act of visualisation gradually dispels that darkness, replacing the void, the space, the unboundedness with human classificatory and bordered domination. In a place in which ice age follows ice age the human senses of sight and hearing intimate an unbounded domination of the earth:

> …as long as the human ear can hear the breaking of waves over deep seas, as long as the human eye can follow the track of the northern lights over silent snow-fields, as long as human thought seeks distant worlds in infinite space, so long will the fascination of the unknown carry the human mind forward and upward.
>
> (Nansen 1911a: 6).

Nansen's profound historical work is one of intense scholarship for its period. It archives the entirety of the 'northern' expeditions from the classical period onwards. He notes that Strabo collected accounts that the *Icthyophagi* (fish eaters) in the far north used whale bones for beams and rafters in their huts (1911b: 160) as well as the lore of the Hyperboreal regions in classical geography and cosmology. Perhaps unsurprisingly Nansen makes much of the Norse discoveries of Greenland and the Icelandic seas with the sailors plotting their routes and locations by examining the routes of whales (1911a: 251).

Nansen recounts the tales of Arab travellers and their accounts of whales (1911b: 145) with chroniclers documenting the captured and flensed whale carcases with flesh as white as snow and skin as black as ink (1911b: 157). Like other archivists of the trade, including the elder and younger Scoresby, Nansen revels in the visualisation of whale vivisection. His text includes copies of medieval pictures of whale vivisection and flensing from the 'Icelandic Land Law'. The whale is tiny in comparison to its three human flensers, each with a different type of knife, with the scythe of one attending to cutting the whale in half (1911b: 155). The human perspectivalism that diminishes the whale to a fraction of its actual size also increases the visual power of the human flensers. Each having a different knife in order to dismantle and disarticulate the whale they wait their turn and their role to practice upon the carcase. Not only do the texts of both Nansen and Scoresby document the parasites and the bacteria of the living whale and the whale corpse specifically as it decomposes, so bacteria itself is used as prosthetic weaponry by the whalers.

Noting that the different practices of hunting, herding and interception entail different methodologies of destruction (1911b: 155) Nansen then discusses historical and current practices in Norway, specifically the Bergen area:

> When the great whales enter certain fjords which have a narrow inlet, their escape is cut off by nets, and they are shot with poisoned arrows from bows which entirely resemble the crossbows of the Middle Ages. The arrows used are old and rusty, and convey bacteria from one whale to another. When the whale has been hit by these arrows it is rapidly weakened from blood-poisoning, so that it may easily be harpooned and then killed by lances, after which it is cut up and divided among the inhabitants of the fjord, according to ancient, unwritten rules. In spite of the blood-poisoning, the whale's flesh and blubber are eaten, and are regarded as very valuable provisions. I have myself often taken part in this kind of whaling.
>
> (Nansen 1911b: 157)

Historically, iron spears were infected by the bacteria from captive or dead whales (1911b: 158). The use of bacterial methods of destruction would prefigure the human genocides and methods of the twentieth century but also points to the uses of cleansing materials and the ideological justification for destroying human germs and their carriers.

The way in which the spear and the lance 'encounter' and 'contact' and indeed pierce the whale's skin can also illustrate some problems in the social memory of the hunters and their social formations as well as the meaning of a whale 'archive'. Scoresby the younger recounts several stories of harpoons and lances that were found present in the whale skin before they were finally made captive and destroyed by the European fleets. The whales had clearly survived previous attacks. What is remarkable about these accounts is the disjunction between different whaling 'civilisations'. Indeed, it demonstrates a clash of human and geological epochs over the body of the whale. Whales captured near Spitzbergen displayed 'stone age' harpoons enmeshed in their blubber (Scoresby 1969a: 10–11). Scoresby notes that 'A small white scar on the skin of the whale, alone marked the place where the lance had entered' (1969a: 10). Another whale was found to contain within its skin a bone harpoon and another a stone weapon. The 'stone-lances and bone-harpoons' entangled in the whale's shell hinted at the intervention of other humans other than the European whalers probably of the Esquimaux residing in a different historic epoch to that of the 'modern' whalers but one historically parallel to it (1969b: 11, and see also Hoare 2009: 292–293).

Scoresby as a scientist, of trans-polar passage, was the first to uncover the complexity of snowflakes, of different types of ice, and of the variegated colours of water (with the whales favouring green) (1969b: 176–177). But he remained obsessed by the refracted mirages that he found upon the ocean and the polar ice. It was the accounts of these mirages in Scoresby that would be hunted out and read many years later and which, as already recounted, would become part of H.P. Lovecraft's Cthulhu mythos. But it also hints at the anthropomorphic translation of barren natural ice into human encampment and city. Scoresby recounts these magical visions:

> Presently the scene was changed; the mountains along the whole coast, assumed the most fantastic forms; the appearance of castles with lofty spires, towers and battlements, would, in a few minutes, be converted into a vast arch or romantic bridge. These varied and sometimes beautiful metamorphoses, naturally suggested the reality of fairy descriptions; for the air was perfectly transparent, the contrast of snow and rocks was quite distinct, even in the substance of the most uncommon phantasms, though examined with a powerful telescope and every object seemed to possess every possible stability.
>
> (Scoresby 1969b: 386)

Translated into the Antarctic south, these visions would become the basis for the vast city of the 'old ones' in Lovecraft's 'mountains of madness' (Lovecraft 2007). The 'fairy' visions of the magical city that Scoresby witnesses is later transformed by Nansen into a vision of the 'fairy north' (1911a: 1). In many ways, the 'never never lands' of the classical imagination would find their nemesis in the northern seas (Hudson 2016a).

The archives of Nansen and Scoresby are archives of destruction. Scoresby himself both literally and indeed metaphorically states of the whales that 'They have no voice' (1969a: 465). We will return to the possibilities of Cetacean memory in due course. Both Nansen and Scoresby also reflect on a central cosmological and ethical document of the medieval Norse period – the 'Konungs skuggsjá' or 'King's Mirror' (Nansen 1911a: 3, 272–272, 1911b: 155 and Scoresby 1969b: 11). Nansen notes the extensive taxonomic Cetacean knowledge of the manuscript (1911b: 155) which is itself comprised of sections on natural history, society and kingship as well as advice on mercantilism and exploiting the natural world. What Scoresby calls the different whale 'tribes' of the Icelandic sea (1969b: 211) are documented in the 'King's Mirror' not only to assess and then utilise them, but also to reflect the natural order of God as expressed through kingship and dominion. Sverre Bagge's commentaries on the 'Konungs skuggsjá' stress the relationship between the natural histories and the social formations of the medieval period (1987, 1994). In his analysis of that dual relation of nature and the social, Bagge rethinks the categories of force, spirits, nature and the supernatural specifically with regard to the metaphor of the *speculum* (1994: 26). The *speculum* is itself the mirror of all things, the accumulation of encyclopaedic knowledge within the bound pages of a book. The 'King's Mirror' itself surveys the seas of its animal and human subjects. Bagge argues that the 'King's Mirror' creates a sociological model of the body politic with God at the head of all strata, hierarchy, and social governance (1994: 32–33). There is a close relationship between the descriptions of the order of nature in the first part of the book on natural history and cosmology and the orders of the social in the second and third parts of the book (1994: 40).

In the 'King's Mirror' the taxonomic and physical servitude of nature parallels the terms of governance of social forms. Disruptions of nature are also paralleled with the disruption of social forms. Indeed, that servitude may only be overthrown by the disruption from within of those taxonomic systems. Sheer monstrousness and hybridities which disrupt 'stable' entities may only serve to be prostrated before new forms of taxonomic description. Perhaps the social memory and archives of the whale are about evading the human bestiary and its descriptions or reasserting new forms of human archiving from below which witnesses to the counter-bestiary of Scoresby's bear in the Tower. The refutation of human description and ordering thereby refutes and renounces subjecthood and physical domination before kings and ship captains.

Cetacean memory

The deposition places of the disjecta membra of the animal dead are perhaps the spaces with which to begin to understand how the 'social' memory of animals can be understood through their material remnants. But these locations are often profoundly different in terms of differing modes of deposition. As the bodies become literally disarticulated they become part of culturally ordered practices. Like human remnants the remnants of animals can be subject to a wide array of practices often resting upon the social and economic utility of their remains. These then can become part of sets of sedimented, stratified systems of depositions which might be understood through excavation. Again, if we examine sites such as Star Carr then we can see, in the excavation and analysis of the material remnants of animal bodies, what these remnants might signify. But critically, their signification processes rest upon what the animal remnants might tell us about the social practices and meaning for humans, not for the animals themselves. The direction of cut marks upon bones are essentially about human utilities and rituals practised upon the animal body. Indeed, the significance for Star Carr might lie in the 'corporeal transformation' of human bodies by the adoption of prosthetic animal remnants.

In one sense, cut marks on bone display what Stiegler has called the 'third memory' of technology and technique: the weaponised material forces brought to bear upon the organic living or dead animal being (or indeed other human beings). Indeed, what we might call the organic bio-memory can often only be reinterred through an understanding of the preserved skeletal remains of the animal whether ossified, fossilised, or petrified. This process of fossilisation and petrification could mark long durational process of deposition and discovery but the ossification process of bone-hardening obviously takes place within the life of the entity and has a limited existence, unless fossilised or petrified after death. The deposition of animal bones tends to be less focused on ritual occasions (although often the use of animal bones has a significant ritual focus in terms of ceremony and sacrifice) and more focused simply on the disposal of carcases subject to disease after the useful material remnants have been taken and used. This in some ways dispenses with the notion of mortuary, reliquary and ossuary processes comparable to those of human beings. The sheep carcase would rarely be perceived in mortuary terms, or as a holy or ritual relic, or collected into an ossuary. The depository of the animal is simply a dump, a boneyard or pyre unless its bones form a decorative, ornamental part of a human burial.

The organic bio-memory reveals itself, as we have noted, in its detritus. Often this is because its skeletal remains are so marked by its life, moulded as they are by existence and growth but also in how it was butchered, eaten and disposed of. The bio-memory is frequently revealed by the utility of its remnants. So, as we noted in an earlier chapter, the translation and transformation of the inherent bio-memory of whale bones into women's corsets tells us something about human cultures, but also about the original entity and its modes of existence since its earthly passing. The ubiquity of whale monuments in the Victorian period (largely arches on paths, fields and church entrances made from whale jawbones), marks the ritual commemoration of human destructive and accumulative cultures, and with it too, a genetically inscribed bio-memory of the destroyed and transformed entity. This is undoubtedly enmeshed in taxonomic and classificatory knowledges of other species on the part of humans, but it can also be the disentanglement of memory from power by somehow restoring a form of memory to the scattered remnant that contests that power and the human projects of whale destruction. It does this by providing a counter-archive to the whalers which challenges their *dominion* by memorialising the whale and its right to exist as a species in the face of human invasion and indeed imperialism. The human empire of butcher and slaughterhouse and the official memory of the trade become contested by the counter-narratives which displace the automatic right and legitimacy of human dominion. It is ironic that Lieutenant Du Roi, sailing on the Whitby-built *Prince of Wales* noted that the front parts of the ships were larger and more bulbous than the rear – this was taken from the experience of 'nature' and of observing Cetaceans, specifically the speed of dolphins (Baines 2010: 72). Like the corset the whale ship itself is modelled on its prey.

In the structure and the lattice-work of the human corset we begin to delineate a history of human destruction and oppression and by necessity a challenge to it. In other words, tracking the animals moves us beyond hunting along the whale routes and deer tracks in order to archive the experience of the animals themselves and to initiate new ways of cohabiting and coexisting.

The social memory of the abattoir, the stockyard, the shambles and the charcuterie, indeed the whole ideology of carnism and the carnage it inflicted finds a counterpart in new archives which reveal and display: not in order to vivisect but to restore, to incarnate rather than disincarnate.

Yet enmeshed in those contested social memory structures is something of profound importance. The compulsion to consume whale meat and oil, to create whale-like structures around the human body with whalebones, says much about the phenomenological relationship of humans to other species. The human world wants to taste the animal world and revels in its destruction, to build similar physical structures because that taste and consumption literally swallows the powers and capital of the animal. We reproduce ourselves through consumption and through our appetites. Not only does animal capital in its congealed form of flesh reproduce our bodies it also reproduces values and characteristics of the animal as we eat it. Often this is simply about physical and biological reproduction but it can also, as meals often do, entail a ritual function of incorporation and the visualisation of social power. Instead of ossifying and rigidifying and petrifying into stone, consumption sublimates another's biological being into ours. However, like the whalebone arches, we also petrify and memorialise as we consume, transforming animals into representations and statues. Substantiating, literally, the social memory of their destruction into ornament and decoration often by carving their images out of the bone of the animal that the artist is attempting to represent. Can we make the whale speak for itself and not for us? Is the whale silent with its own memories?

The entanglement and mutual comprehension and incomprehension between the whale and the human being is most remarkably expressed in Laurens van der Post's book *The Hunter and the Whale*. The novel is based on his experiences of the South Atlantic whaling industry and continues his reflections on hunting, animals and humanness that would culminate in his work on the Kalahari. Initially, van der Post distinguishes between the bounded and fortified camps and civilisations of humans and the boundless 'unknown' nature beyond it. Nature is 'uncircumscribed' by humans (1970: 11–12). The encounter between the animal and the human for the first time often initiated new forms of knowledge, lore and superstition including ideas of the dead animal entering a human soul as they smell the dead corpse (1970: 12). The 'column of smell' stands above the whale and above the ocean reminiscent, for van der Post of the smell of the dead on his own human battlefields (1970: 106). Van der Post's account is profoundly anthropomorphic. Rather than seeing a relation, compatibility or cohabitation with the whale, he, as a hunter, imposes human qualities of love and freedom upon the vast mammal (1970: 109–110). This is in the midst of the 'mechanical' destruction of the 'internally lacerated whale' (1970: 111). The whale corpse is then photographed (1970: 115). One day the whale ship passes a shoal of killer whales and the whalers see their own appetite for death and destruction mirrored in the other species. Their feeling is both of awe and self-hatred as the whales pass (1970: 139–140). Yet in the next moment van der Post revels in the accumulative power of the whale ship and the hunter. He reveals the appetite for spermaceti, the substance taken from the head of the whale and used for oil and lubricants; he passionately details the love for the sperm whale's faecal ambergris (1970: 141) – its musky scent used primarily in perfumes that might bedeck the breasts of humans, perhaps as they themselves are wrapped in their whale corsets – not only in order to 'disguise' themselves as whales but to subsume the power of the mammal into their social orders, ornamentation and fashions.

For van der Post the phenomenological relationship between the hunter and the hunted is expressed most forcefully in the contest between the great sperm whales, most often found in the teeth of storms and cyclones, and the whale ships (1970: 142–143). The great, lone 'l'Empereur' or 'Caesar' finds its nemesis in the whale ship hunting it for its spermaceti and ambergris (1970: 143). Yet for van der Post, the gradual mechanical distancing of the whale destruction, separating each contesting species out through prosthetic machines decreases the lore and knowledge of the whale on the part of the whalers (1970: 144). The vastness and scale of natural destruction exemplified by both the hunter and the hunted is also reflected in the evolutionary cycles of departure and return of the whale: not just in terms of its geographical routes but also its evolutionary routes. The passage of the great mammal from the land into the sea, bearing the characteristics of its origins still, itself reflects the journeys of human beings between the land and the sea (1970: 147–148). In itself this is a biological and technological revolution internal to the whale anatomy. Humans see themselves in the depths of the whale, in their evolution and anatomy. Jonah could only become a human prophet because he had once been inside of and part of the whale (1970: 149–150).

But still the whale does not speak for itself and for its social memory as a species. Can we evade the translation and the intercession of humans as we ourselves are humans? How can we focus on the whale world and not the human world? And are those worlds still in co–existence as two or more states of phenomenological being each traversing the seas? George Steiner in his work on language and animals is one of the few to address what he calls the 'expressive codes' of other species. For Steiner,

> Life proceeds amid an incessant network of signals. To survive is to receive a sufficient number of such signals, to sort out from the random flux those literally vital to oneself and one's species, and to decode the pertinent signals with sufficient speed and accuracy.
>
> (Steiner 1972: 58)

The signals of animals are statements of ontological existence, even of communication: 'Only the inert is mute. Only total death has no statement to make' (1972: 59–60). The end of human civilisation and even of the human species itself will not terminate signals and codes of expression. As Steiner argues:

> A world without words can be, and, where organic forms are present, must be, a world full of messages. Language is only one, and probably the most recent, of a great sum of expressive codes. Not only do these other codes persist; they may well outlive language. A post-human planet, so long as zoological phenomena endure, will teem with significant, conventionalized communication, as did the earth in the Paleozoic. After man, there will not be silence.
>
> (Steiner 1972: 60)

The post-human world and the species that persist beyond it will not, as Steiner says, be silent. So how far can the expressive codes and signals of animals and specifically whales encode the social memory of those species? And how do they address us?

Seeing and listening to animals

If we can at first examine how humans 'capture' or 'make captive' the *human* memory of animals. Like the photograph of the dead whale that van der Post notes above, the capturing of a photographic moment, witnesses to something: even if that something is the destruction of an animal. The official archives of stockyards and butchers often contain within them certain types of counter-archives and counter-narratives. There is no necessary collision between the use of official and unofficial archives and narratives if approached with caution (see Hudson 2016b).

If we return to parallels with the Jewish Holocaust, we can examine Gigliotti's analysis of the cattle-trucks again (2009). For Gigliotti,

> Photography, like cinema, provided a screening of deportation's procedures that suggested the trauma of victims. These depictions were limited to exterior depictions of the train and its passengers at departure, in transit, and at arrival...Cinema's intervention was to recreate the camps and their death-world as deportation's destination, locations largely omitted from the historical visual archive of resettlement.
>
> (Gigliotti 2009: 12)

Nobody would deny the usefulness of the photograph and the film of the trucks for understanding trauma but it provides, as Gigliotti says, an exterior depiction of the cattle-car experience. But there are frames that are not seen, interiors that are not witnessed, counter-archives that do not emerge. For animals, how do we begin to delineate this interior world, beyond the official photographs taken by the hunters? John Berger begins to elucidate the different temporal and visual rhythms of the animal archive. For Berger:

> The speed of a cinema film is 25 frames per second. God knows how many frames per second flicker past in our daily perception. But it is as if,

> at the brief moments I'm talking about, suddenly and disconcertingly we see *between* two frames. We come upon a part of the visible which wasn't destined for us. Perhaps it was destined for night-birds, reindeer, ferrets, eels, whales…
>
> (Berger 2001: 5)

Indeed, the evasion of 'capture' lies in the interstitial, the ellipses between the frames, between the human shot and the human intention. The animal reveals itself when we are not looking – its images not *destined* for us or to be ours. Perhaps this hints at an existence without our intercession or translation: simply that other beings and species co-habit with us in our world. Berger notes this moment of coexistence as day breaks over the earth: 'In the darkness, which precedes sight or place or name, man and calf waited' (1979: 21).

The anthropological definition of humans and their relations to other species like Orang-utans and other parts of humanity like feral children (Barnard 2000: 18–19) must be revised in terms of those relations. That these other beings and entities are not necessarily exclusive to us, separate, or beyond the human. Indeed, the reassertion not of their anthropomorphism but of our animality may lead to new modes of coexistence violated in the metabolic rift of the human invasion; those genetic inscriptions of animality may also be violated and liberated through technology and the supersession of a social memory resting upon biology and the organic. Bernard Stiegler announces this violation:

> It is by freeing itself from genetic inscription that memory at once pursues the process of liberation and inscribes thereupon the mark of a rupture—on stones, walls, books, machines, madeleines, and all forms of supports, from the tattooed body itself to instrumentalized genetic memories, disorganized, made inert [*inertifées*] as it were, then reorganized, manipulated, stored, rationalized, and exploited by the life industries named "biotechnologies", including the holographic memories that the information-processing industry is planning.
>
> (Stiegler 1998: 169)

The rupture between the organic and mechanical memory through technological revolution itself echoes the revolution which overturned the original balance and coexistence between species. The way we look at and examine animals in that process changed radically. For John Berger:

> The 19th century, in western Europe and North America, saw the beginning of a process, today being completed by 20th century corporate capitalism, by which every tradition which has previously been mediated between man and nature was broken. Before this rupture, animals constituted the first circle of what surrounded man. Perhaps

> that already suggests too great a distance. They were with man at the centre of his world.
>
> (Berger 1980: 1)

As we noted, even the distance of the hunter and the hunted through machinic transformation was itself extended. The relationship between the human and the animal was itself a social relationship based around their imagination of one another: their comprehension and incomprehension (1980: 4). This is not just the human speaking for the animal, imagining the animal and translating it, but an intercession into human nature and its origin in animality. The imagining of the animal is an imagining of animal origins. As Berger argues:

> In one sense the whole of anthropology, concerned with the passage from nature to culture, is an answer to that question. But there is also a general answer. All the secrets were about animals as an *intercession* between man and his origin.
>
> (Berger 1980: 4)

The metaphoric relation between human and animal is reaffirmed by the fact that the animal was the first and originary metaphor that humans used to understand themselves (1980: 4).

But what about the animal imagining us? As Berger notes:

> animals are always the observed. The fact that they can observe us has lost all significance. They are the objects of our ever-extending knowledge. What we know about them is an index of our power, and thus an index of what separates us from them. The more we know, the further away they are.
>
> (Berger 1980: 14)

Taxonomic and classificatory systems of knowledge have separated them as a set of zoological species themselves to be observed most effectively in the animal camps of the carceral zoo system (1980: 22). Like Scoresby's bear, torn from its habitat and taken to a bestiary, the zoological visualisation of animals severs, for Berger, their relationship with other species and subordinates them to a central relationship with their 'keepers'. The distancing of the animal from the world that was theirs, its geography and its other species, relocates it in a cage to be observed (1980: 22–23). For Berger, not only is the zoo a site of 'enforced marginalisation' – a parallel of the 'madhouse' and the concentration camp, it is also a space of extinction. For Berger: 'Everywhere animals disappear. In zoos they constitute the living monument to their own disappearance' (1980: 24). They disappear even as we are watching them.

Rethinking dominion

David Rothenberg has asked: 'Why would a whale care about human sound?' (2008: 10). It may be that the entirely different universes of these species make communication difficult if not impossible, but the domination of the whale empires by the human empires illustrate not just physical but cultural dominion. Domination is often about the entangled dialectic of knowledge and suppression: humans observe in order to dominate. But there may be ways that we can rethink the human relationship to other sentient species by thinking about the religious conception of dominion in which human dominion entails the servitude of humans before nature. Indeed, dominion does not necessarily entail human despotism. Andrew Linzey's theological work has attempted to think about dominion in terms of care and love for 'creation' (2007). Developing new ways of moving beyond a liberation theology which privileges 'humanness' Linzey and his collaborators have redefined the differing estates of human and animal (Linzey and Cohn-Sherbok 1997). Animals are not just there for the use of humans but to manifest creation itself (Linzey 1994, Linzey 1991).

Critiquing the humanism of Liberation theology, and specifically Guttierez's version of it, Linzey dispels its obsession with human domination:

> Liberation concerns oppressed "peoples", "humankind" growing in historical consciousness; the realisation of the dimension of "humanness"; Christ indeed is defined as the one who "makes humankind truly free". In short: the first major work of liberation theology espouses an uncompromisingly dogmatic humanocentricity.
>
> (Linzey 1993: 508)

The idea of the 'humanocentric' world must be displaced by a concept of all sentient 'things' as subjects of creation and suffering (1993: 509). The co-existence, before God, of the human and animal replaces the kind of dominion that the elder Scoresby wielded over his captive bears and whales, imprisoning and vivisecting them as we have seen. Equally compelled by a vision of God a new animal theology would involve new forms of co-habitation between the animal and the human as sentient beings of salvation. For Linzey:

> Another way forward could be to show how a frequently used concept like "neighbour" includes what its biblical sense intends, namely those who exist *alongside us*, and how it is not infrequently that non-human animals are our neighbours, or more precisely our conscripted companions, in this sense.
>
> (Linzey 1993: 514)

This entails the dramatic transformation of both the concept of sentient creation and the divine right of human beings to dominate. It does not mean displacing humans from their position of care and responsibility however. Dominion entails the defence of the weak and the powerless and the ceaseless search for the liberation of all sentient beings who can actively suffer and be oppressed (1993: 525). This dominion equally challenges the idea of ecological management of humans and their capacity to destroy species in the name of 'biodiversity'. Dealing death and destruction to animals perceived to threaten biodiversity has to be replaced as a form of 'hubristic domination' of other species (2001: 21).

Andrew Linzey's theology of human displacement from master to servant rests upon a biblical concept of creation and the reassertion of animal care through recourse to biblical textuality. The contradictions of sacred texts are one thing, yet another is the radically different practical programmes that can ensue from the idea of human dominion including tableaux justifying and displaying the destruction of whales to Linzey's own theological animalism. If sentience entails the endurance of suffering and oppression then how far can we move beyond this anthropocentric empowerment of the weak in order to truly create new coalitions, cohabitations and even coalescences between animals and humans? This might indeed lead to cohabitations which entirely rest upon separatism and separate modes of evolution with different quarters, estates, domains and even oceans demarcated to different species. If the contestation of species has led either, to paraphrase Marx, to the ruin of the contending species or to the supremacy of the human species can this end only by re-bordering the human-animal world in terms of, for example, processes of re-wilding rather than ecological management and conservation?

Often political prophecy in the medieval period was demonstrated in 'Galfridian' terms – using animals as a form of disguise for discontents and satirists (see Taylor 1911). Often these satires, directed as they were against the monarchy and the church, created new kinds of parliaments, commons and estates of animals. Excluded from the social estates of the medieval period of king, noble and merchant, both peasant and animal were represented as estateless. They owned nothing and had no recourse to anything but the common land that they shared with their human and animal kin. If we see, as Linzey does, that human dominion means being a servant to the animal species in their care, this further entails some sense of retaining, archiving and caretaking by the human species over others. Still others have argued that the human invasion of the animal world should essentially be rolled back and human dominion curtailed by radical population reduction moving humans back to 'pre-invasion' boundaries (Lee 1990). The moral ethical imperative to self-eliminate a species because it is, in essence, too destructive would mean the complete supersession of the human world to ecologies that were no longer managed and not even re-wilded. But it may be that human cultural and ethical transformation could make new compacts, treaties and coalitions with other species based on the construction of *incompatible* life-worlds, domains and estates. The co-habitation of spaces of coexistence could continue but at the same time spaces could be constructed like the 'empty' republic of Antarctica which is purposely as well as climatologically bereft of human species and intervention. The human disposition towards violence and subjection could be entirely eradicated from specific non-human encampments. This may also help understand the prefiguration of human violence in prior forms of violence against animals. It may be that new forms of ultra-humanism may indeed support the creation of new kinds of relationships of liberation with other species; of restoration, of new forms of ecological and sustainable settlement and perhaps even new worlds entirely (Armstrong 2014, 2016). It may be that the vision of vast industrial cities is replaced by new forms of mobile encampment which are more akin to ecologically balanced prehistoric social formations which might no longer signal human interventions into geological strata. Indeed, these counter-anthropocene communities could become new models for afforestation, new oceanic strategies of non-intervention, new forms of civil society and commons in which animals are liberated and empty quarters and republics which are non-human spaces.

Biology, species and memory

Is there such a thing as the 'agency' of whales? What is the relationship between whales as material entities and discourse? For D. Graham Burnett this epistemological quandary is central to his own archiving processes: 'Do I give the whales "agency" in this book? Not really. There aren't any whales in this book, remember? Only words about whales', (2012: 9). In this section we return again to Cetacean memory and specifically its biological foundations in order to think about human 'exceptionalism' and evolutionary convergences between species around memory, self-recognition and sociality. Jason Bruck in his study of memory in dolphin Cetaceans has argued that long durational memory structures support evolutionary advantage around reproduction (including memory so as not to inbreed), and remembrances of past interactions. He argues that this is necessary for the social existence and 'reciprocal altruism' of the whale group and its survival in the face of social threats (2013: 1). Like humans and possibly elephants the Cetaceans have a capacity over long period of time for social recognition and complex and fluid socialities. These fluid and complex socialities, he argues, may in turn propel and promote cognitive development (2013: 4). Bruck argues that the relation between fluid and complex sociality and cognitive development is not definitively proved by his data and study (2013: 4) but the interaction between social form and encephalisation has been noted in other studies of Cetaceans. For Bruck, the implications of his study intimate an important set of connections between memory and social formations of Cetaceans:

> We need to further address animal long-term memory to see whether non-social information will also show the same level of resilience or whether social content is more salient and therefore more resistant to

> decay. The broad implications of this work, however, suggest that at least social memory in non-human animals is perhaps more resilient than previously thought, as no systematic study so far to my knowledge has shown such long retention of information in a species outside of humans.
> (Bruck 2013: 4)

Other studies of memory in African elephants and migratory songbirds respectively (McComb et al. 2001 and Godard 1991) have focused on long-term memory structures with the remembrance of certain sonorities or the retention of social knowledge in the memory 'depository' of the elephant matriarchs (McComb et al. 2001).

The remarkable work by Janet Mann and her colleagues on Cetacean social structures has transformed our notion of sentience and cognition in dolphins and whales (2000). They note that the new analyses of Cetacean social structure had their origins in previous generations of primatology and were indeed inspired by new research on primate social formations (2000: 1). They stress in their introduction to this set of field studies the importance of comparative taxonomies of Cetaceans with other mammals including humans, the use of theoretical and ecological frameworks of cetology rather than simply data-driven research and the necessity of long-duration observational research into the socialities and social structures of Cetaceans (2000: 2). They also highlight the recurring and possibly one of the most fractious issues in terms of Cetacean research and data: the inherent problem of observation. Should cetologists work on carcases and the physical remnants of dead Cetacean bodies, observe Cetaceans in captivity, or study them in their naturally occurring habitats (2000: 2)?

Samuels and Tyack, in their history of the study of Cetacean societies, have argued that Cetacean studies had reached a pivotal juncture by the millennium between natural-historical qualitative approaches based upon accounts and more focused quantitative analyses with the development of more 'Cetacean-centric' methodological approaches rather than within general studies of mammals (2000: 9). Observing and indeed archiving Cetacean sociality was made difficult precisely because of its oceanic and abyssal status. Samuels and Tyack note that: 'The sea is additionally hostile to tangible signs of lifestyle such as tracks, scats, or nests – a factor with which few land-based biologists have had to contend', (2000: 10). Therefore the 'decoding' of Cetacean social systems had historically been left to what they call the 'whaler-naturalists' (2000: 10–11), whalers at once observing and then destroying their prey. For Samuels and Tyack:

> …for centuries, human interest in large cetaceans has been a commercial one. Through the whaling eras of open, human-propelled boats and sailing ships to the modern factory vessels, any knowledge acquired about the social behaviour of the great whales was employed to increase whaling harvests or to manage a lucrative but diminishing resource. As a result of

> this exploitative association, knowledge about whale social behaviour has been built upon foundations of population biology, stock management, and analysis of whale carcases and whaling statistics – avenues of inquiry whose methods, philosophies, and vocabularies imposed idiosyncratic ways of thinking about whale behaviour.
>
> (Samuels and Tyack 2000: 10)

The displacement of 'whaler-naturalism' by more conservation oriented approaches was also paralleled by the increased mechanisation of whaling which distantiated the observer from the mammals and the process of their destruction (2000: 17). But the history of Cetacean studies was still bedevilled by the methodologies of observation. Making deductions about social formation and structure from the physical remains of the Cetacean was similar to making paleontological deductions about 'never-seen, extinct animals' (2000: 20). As Samuels and Tyack say, there were significant problems of 'Inferring social behavior from dead bodies' (2000: 21). At the same time studies in captive settings were raising ethical and methodological issues (2000: 22). The emergence of observational and non-invasive field studies they argue issues in new methodological descriptions of naturally occurring whale activity in their habitats (2000: 32–34).

New methodologies for understanding Cetacean social structure began to emerge which privileged the idea of memory and social structures that persisted over long periods of time and across oceanic geographies (Whitehead, Christal, and Tyack 2000). Field studies also led to a more nuanced understanding of functionality in terms of memory and communication across oceanic space (Tyack 2000). With these new non-captive spaces and a focus on conservation, new compacts and forms of management have begun to emerge which privileges the rights of Cetacean species (Whitehead, Reeves, and Tyack, 2000). The new focus on conservation has robustly challenged the concept of the whale as a form of capital, social utility or accumulative resource. Jamieson and Regan have documented the ways in which whales specifically should be understand as having analogous rights to those of humans and refute the whale as an object of consumption. Further, they argue that the whale should not be there for observers to study, watch or 'play with'. We should leave the whale be in its own habitats and evolutionary processes (Jamieson and Regan 1984).

Lisa Marino has led the ethical programme for the non-human rights of Cetaceans against those who still hold to the methods of 'captive studies'. Marino and her co-thinkers argue for less or non-invasive observations of Cetaceans. The scientific limitations in terms of knowledge are obvious and there are serious considerations about the stress on the Cetaceans themselves (Marino and Frohoff 2011). Instead, Marino and Frohoff propose something startling: the new approach of *Interspecies Collaborative Research* which supports the autonomy of Cetaceans at the same time as bringing together humans and Cetaceans in naturally occurring conditions. It may be then that Cetaceans themselves might be the recipients of the benefits of new types of scientific knowledge that might emerge co-produced by humans and the mammals themselves (Marino and Frohoff 2011). These new kinds of coalitions and alliances may indeed echo the kinds of alliances forged within and between different oceanic species (see Connor, Smolker and Richards 1992). It may be that these new understandings might displace human 'exceptionalism and primacy' and as Rollin has said, create a new era of human–animal relations. This may displace the understanding that we have had of animals since the ancient classical period (see Hudson 2016a). For Rollin: 'Only when our actual decisions and actions reflect a moral regard for other creatures can we truly be said to have escaped the stranglehold in which the conceptual scheme of human separateness from nature has held us since antiquity', (Rollin 1981: 79 and see Bekoff 2003).

Lisa Marino has argued for the emergence of convergent intelligence in both cetacea and primates specifically with regard to communication and self-recognition. Indeed, she argues for the idea of 'Non-human culture' supported matrilineally within sperm whales. The primate–Cetacean emergence of a parallel intelligence to humans display that very different evolutionary routes can converge in similar levels of intellectual and potentially cultural capacity (2002). Perhaps the greatest documentation of the sperm whale is that by Hal Whitehead and his colleagues specifically in terms of the social evolution of the whale and its companions (Whitehead 2003).

In some ways, the convergence of intelligence and social structure of the human and the whale can be illustrated by the sonorities of each species. The 'philosopher-naturalist' David Rothenberg in his analysis of Cetacean–human sonorities has noted that: 'Maybe music is a part of nature itself, something evolution has produced on different lines, converging into some living beauty that whales, birds, and even humans can know', (2008: 6). The sonic interactivities of human and whale, even though living in different worlds (2008: 8), could potentially create new modes of coexistence, convergence, reparation and restoration. Potentially the differing metabolisms of the human and whale, hitherto in rift, could support new sonorous encounters of mutuality between species.

Archives of cetacea

> The analysis of nature into its individual parts, the grouping of the different natural processes and natural objects in definite classes, the study of the internal anatomy of organic bodies in their manifold forms – these were the fundamental conditions of the gigantic studies in our knowledge of Nature which have been made during the last four hundred years. But this method of investigation has also left us as a legacy the habit of observing natural objects and natural processes in their isolation, detached from the whole vast interconnectedness of things; and therefore not in their motion, but in their repose; not as essentially changing, but as fixed constants; not in their life, but in their death.
>
> (Engels 1934: 27)

The assault on the 'lifeless positivism' of mechanical analogies would lead Engels towards a dynamic understanding of nature (Hudson 2000: 265). This new 'dialectics of nature' would in turn be debased and used for totalitarian purposes (Lecourt 1977). The question of whether nature was 'dialectical' and the whole question of dialectical and historical materialism would continue to bedevil Marxist theory for the whole of the next century and beyond. Yet Engels' understanding of methodologies of observing nature is still pertinent. The epistemological shift was between two divergent methodologies: one of stasis and the other of movement. The classificatory, taxonomic impulse was one which categorised, visualised and separated: whether natural object or living species. It examined species; birds, fish, Cetacean in their death as objects of analysis. As Engels says: an examination of an object in isolation, detached, disconnected. Indeed, the examination of the internal anatomy of an animal could only rest upon the dissected corpse of a dead, disconnected creature. This was the observation of a 'fixed constant', an animal in repose, silenced by death. Yet the mobile materialism that Engels proposes sees the being as connected, in movement, living. Far from observing the 'instant' in isolation and disconnection Engels begins to see the instant as simply the instantiation of a vast network of instant, connections and relations – of different entities and different forces. This then displaces the 'constant' with the endlessly shifting, transforming, replacing and recomposing. The reproduction and the sociality of entire species of human and Cetacean can only be properly observed in the 'dialectics' of biological existence and mobility.

The methodologies of species observation and description that we noted above have elaborated upon perceiving species in repose, in captivity and in their natural habitats. Yet the mobility and migratory movements of the whale, navigated as Philip Hoare has said 'via age-old and invisible signs' (2009: 24) are perhaps the truest explication and elaboration of the method that Engels proposes. The captive whales of the aquaria of the American and European cities were imprisoned in tanks which could literally not house their vast mammality and what Hoare calls their 'supernatural physicality' (2009: 30). For Hoare this 'landless race' of whales

> defy gravity, occupy other dimensions; they live in a medium that would overwhelm us, and which far exceeds our own earthly sway. They are Linnaean-classified aliens following invisible magnetic fields, seeing through sound and hearing through their bodies, moving through a world we know nothing about.
>
> (Hoare 2009: 30)

Indeed, our entire industrial modernity is based on the suppression of whale mobility and survival and their near-extinction (2009: 38). There was a convergence of the whale and the slave trade: ships used for slaves became whalers, freed slaves became tenured labour on those whale ships. As Hoare notes: 'They resemble nothing so much as the slave ships of the age: the one designed to carry the harvest of dead whales; the other to convey living souls', (2009: 101). The exploitative 'transoceanic' industries of slavery and whaling, for Hoare, co-existed as accumulative and extractive mercantilism (2009: 104). As Nathaniel Philbrick has noted – by 1760 almost all the whales around Nantucket had been driven to extinction by hunting so the circles of extinction were widened out into the Atlantic and Pacific worlds (2000: 6). The factory-ships of the whale hunt would follow the mobile migratory movements of their prey, learning their ways and attempting to understand their cosmology. For Philbrick: 'Just as the skinned corpses of buffaloes would soon dot the prairies of the American West, so did the headless gray remains of sperm whales litter the Pacific Ocean in the early nineteenth century', (2000: 65).

The marvellous analytical work of D. Graham Burnett has furnished a new understanding of the search for 'whale knowledge' (2012: xvii). His analysis of the great New York whale trial was a historical analysis of epistemological struggles over the order of nature and of the whale as fish or mammal (2007). Burnett's grand analytical study of twentieth-century Cetacean science is unparalleled (2012). Ostensibly a history of Cetacean science it notates the archival history of 'sloppy slaughterhouse anatomy conducted under macabre and trying conditions…' and of hunting and science in the 'fantastically lucrative circum-Antarctic killing fields' (2012: 3–4). The industrial-scale cetological carcase studies of Antarctica are surveyed (2012: 63–64). The process of Antarctic flensing which takes place in the name of scientific advancement (2012: 66) takes place in the same location and with the same levels of butchery as that documented in the camps of science by Lovecraft below the 'mountains of madness' (Lovecraft 2007). Burnett, examining a whole century of Cetacean scientific practice, begins with the *Discovery* Investigations of the 1920s and the relentless destruction of whales in order to anatomize them and to gain 'anatomical indices' of whale knowledge (2012: 57). Uncovering the role of labour and experimentation over the whales in the Antarctic 'flensing camps' illustrates for Burnett the sheer number of specimens collected and destroyed (2012: 64). This was a shift from understanding an individual of a species to understanding whole populations. As Burnett notes: 'One could begin to think about a given whale as a sample of a population, rather than as a carnival singularity washed onto the beach' (2012: 64). The Cetacean became a phenomena, an object of enquiry. Their skulls mapped (2012: 209), their evolution and 'genealogical sequence' understood (2012: 221). The whale was an object and phenomena to be retrieved, experimented upon, observed, recorded, destroyed and ultimately conserved. By the end of the twentieth century Cetacean science had literally undergone a 'sea-change'. The focus on the dolphin and whale mind, sociality and communication, largely initiated by the work of John C. Lilly and his collaborators, was transforming the Cetacean into an equivalent, as a moral agent, of a human being. For Lilly, the whale mind was to be understood as vast processing computer, an archive of memory and experience that the whale mind could access and use

(1967: 56). As Burnett notes, Lilly was the first to empirically describe the Cetacean as an agent and as a social being. This was done however through often 'vigorous' experimentation and included 'neurological vivisection' (2012: 532). Burnett includes in his text Lilly's own map of the 'animal camp' of the St Thomas Institute including pools, observation decks, photography lab, data-processing labs and a dolphin laboratory under which was a pool from which they were extracted (2012: 584). Ultimately the whales would become, Burnett argues, agents of our own moral consciousness and 'assemblages' of our own human cultural aspirations (2012: 642). The whale emerged as a being of 'extraordinary symbolic power' (2012: 643). For Burnett, the scientific description of the Cetacean dispelled archaic symbolic representations in order to present itself symbolically anew:

> Few disciplines can more straightforwardly tell their story as one of gradual demystification: once whales were *the* great unknown – monsters of the deep, Leviathan, known only to God Himself. Gradually the wild legends, the seafarers' yarns, the biblical tales, all gave way – fell away – as men of learning pursued the cetaceans with harpoons, microphones, nets, and neurological probes, leaving a pole of quaint old beliefs in shards (on the one hand) and a proper knowledge of what cetaceans really were (on the other).
>
> (Burnett 2012: 643)

The science of the Cetacean was still though an archive built by humans of whales and not whales of themselves.

In the belly of the whale

The whalebone corsets which shaped and encircled the bodies of fashionable women created a sense of vastness and of the extension of the female body into a Cetacean-like structure. Like the corset, so the whaleboats themselves were designed to parallel and complement the body of their prey (Ansel 1978). The detritus of the trade were scattered in fields and churchyards where whalebone arches often replaced wood as symbol of power and domination over nature (Redman 2004). Recent archaeological work in the Arctic north has demonstrated the ritual determinants of whale bone assemblages (Whitridge 2002) specifically around spatial structures and locations. The use of whale bone for structural and architectural utility in prehistoric Atlantic Scotland has also raised issues about whether the whales were hunted or stranded (Mulville 2002 and see also Clark 1947). Savelle's work on architectural utility and whale bone assemblages has noted the respective values of whales for meat or for structure. The author argues that architectural utility is the primary factor in the bone assemblages of the Canadian Arctic (1997).

Vicki Szabo's powerful analysis of medieval whaling has argued that the hunt was about the critical subsistence of the North Atlantic peoples (2008: 4).

Szabo notes the relations of ideas about to whale to practices around its utility for the hunters (2008: 4). But there was a central utility in terms of the bones for structures (2008: 6, 206–208). Szabo notes the kinds of classificatory orders of the 'place' of the whale in the natural and human world. For Szabo: 'Perceptions of what whales were, how they behaved, and how they were to be used must be reconstructed with an eye to a multiplicity of traditions. Animals, like nature itself, were not simple constructs born of singular tradition', (2008: 16).

Houses made from bones were often central to North Atlantic cultures. Szabo argues that even when the whale bones had been structured into the shape of a house the house itself retained some of the character of the entity of the whale: its animal form continued in the shape of the house. It was both 'house and creature' in which the dreams of its inhabitants were occupied by visions of the dead whale – specifically of the last tempest which had stranded the whale upon the shore (2005). For Szabo:

> Medieval people most likely would not reject whale bones or flesh derived from a creature of mysterious origin. However, the works of medieval and post-medieval authors reveal the pervasive monstrous qualities believed possessed by whales. Given the inextricability of cultural perceptions and uses of animal resources, we must consider how perceptions of creatures may have impacted use, rather than merely dismissing them as fictions. Throughout the medieval North Atlantic, whale bone served as a simple utilitarian material, used in a wide variety of applications. Given the importance of whales as resources within the medieval world, and their clear impact upon the medieval mind through literary tradition, further thought must be given to the relationship between perception of whales and their utility. While difficult to answer, such questions must be asked in order to appreciate the unique cultural and economic contributions of whales to the medieval world.
>
> (Szabo 2005)

The classificatory power to place the whale in the category of the 'monstrous' coalescing it with fabricated and fictional monsters also reflected the power that would be subsumed into the individual as they consumed it into themselves. The corset, ship and house, in destroying the whale as a living entity, preserved the social and imaginary powers of the whale into new forms of ritual and social practice.

The consumption of the flesh of the whale and the use of its skeletal structure for vernacular architecture displays a dual process. In the first the whale enters the interior of the human being in order for the human beings to satisfy their appetites and in order to reproduce themselves and members of their social group. But the human then enters into the interior of the whale, stripped as it is of edibility, in order to survive in the Atlantic north and to experience a phenomenological or imaginary relationship to the dead entity.

In either case the remnants of the whale become transposed, preserved and elevated into new human, social and architectural forms. The encampments of the human in the north are in the proximity of this other species, utilises that species, and preserve themselves as human habitations because of the whale. The prosthetic utilisation of the skeletal structure as human encampment wraps the human beings inside the 'belly' of the whale and the whale-like replica preserves its form in this new manner. The ossification and petrifaction of the skeletal structure literally humanises the natural form into a mortuary structure that remembers (dreams of) the previous entity as that entity is ultimately hunted into near-extinction.

The distinction between the hunted and the stranded whale is a significant one. The hunt is an active attempt to destroy. The scavenge utilises remnants of an already dead or dying creature. In some ways this reflects the littoral and liminal space between hunting and subsistence agrarianism. That space between the hunted and the stranded whale would be surmounted by North Atlantic peoples in their tool technologies and extended into the practices of the European hunting systems reflecting the distinction between the different epochs of stone/bone and that of iron and steel. Far from the coexisting, cohabiting subsistence cultures of the north the European hunting cultures would create vast networks of industrial-scale whaling often allied to the ongoing use of stranded whales on the European coastlines like the Tunstall whale (Hoare 2009). The knowledges and utilities of the great creature would go hand in hand to extend the observation and more effective hunting of the whale. Taxonomy, classification and human archiving in servitude to the near-extinction of the whale.

7 Natures, cultures, futures

In Chapter 7 we conclude by summarising the discussion and looking at the potential destinations of what I call the 'human movement'; its limitations, migrations, extinctions, crimes and the idea of the restoration of nature and the 'wild'. The potential displacement of the human animal from the pinnacle of the biological and evolutionary hierarchy and indeed the dissolution of that hierarchy has opened up new possibilities for human-animal-nature relations. It also signals a warning that the interventions into nature that began with the Neolithic revolutions (and before that the very birth of the idea of the human in the Palaeolithic caves) and then with the obsession of the classical world with machinery may have to dramatically alter. The maintenance of those systems of metabolic and organic exploitation and the extraction and accumulation of the resources of inorganic nature cannot be sustained. Society entails memory: there are no socialities without forms of remembrance. This is as true for Cetaceans as it is for humans.

Derrida's discussion of the 'Archons' of classical Greece is compelling in its understanding of the relationship between power, knowledge and the law. The Archons hold the depository of documents and guard them. They decide what goes into the archive and who has the right to speak for it and of it. Further, they guard their right to the interpretation of the archive and specifically the documents of law. The Archons speak the law, impose the law and out of the locality and specificity of the document legislate the social order (1996: 2). But who are the Archons of nature and of, specifically, the species subjugated by humans? Can the archives of a subjugated species be guarded, curated and spoken for by the species in contestation with them? Indeed, even within the same species, can contesting forces archive each other or counter-archive each other's histories, documents and identities? The observation of dead, captive or free Cetaceans and the epistemological and ethical effects of this are reflected in the modes of archiving within human communities. If whaling and the slave trade were parallel and often in conjunction how far can we begin to understand the prefigurations and relations between human suppression and whale destruction for example? Observing, speaking for, legislating over is a process of power extracted from knowledge and intimately related to the project of Enlightenment and dominion.

Historically, the Archons of human communities are those who can claim a primary relationship to the land and the earth. In Classical Greece the concept of autocthony expressed this relationship of terrestrial power: those born of the earth, the indigenous peoples sometimes linked to the *Gaia*. The extension of human settlement, imaginaries of ancestry and ideologies of ownership and place were archival gestures used to prove one's lineage as a community in a specific location. The endlessly shifting aquatic/marine archive is profoundly different. It is perceived as unknowable, as a different 'world', as full of monsters. The *terra incognita* of the frozen north, as Nansen has documented, is perceived not just as a place without ownerships but as waste or as commons. Often the bodies of the beasts of the sea were seen as the common right of the hunters of the northern peoples. The abyssal, oceanic history and archive is expressed best by Braudel in his totalising history of the 'middle sea': 'Its character is complex, awkward, and unique. It cannot be contained within our measurements and classifications', (Braudel 1975: 18). Gilroy's 'Black Atlantic' is black and unfathomable in many ways (Gilroy 1993: 16–17, Hudson 2016b: 8). Balasopoulos bears witness to this unusual historiographical status of the sea:

> The sea remains unassimilable to the mainstream historiographical topography of nation and state, capital and province, or metropolis and periphery, just as it persists in defying the striations of borders and boundaries, the furrows of cultivation and civility, the marks of ownership and sovereignty.
>
> (Balasopoulous 2006: 133)

In some ways the recovery of the 'wooden world' of the ship (Rodger 1988: 13) is the discovery of the new world of capitalist modernity itself born in motion. The relationship between the machine of the ship and the birth and reproduction of the 'nautical proletariat' has to be understood in a specific manner: that oceanic history and archiving is profoundly different from 'terrestrial' history. Indeed, if there was an autochthonous species of the ocean, one with a right to speak or sing for itself, it would be the very Cetacean submerged by the human invasion of the marine depths and vastness. The ship did not just mirror the social formations out of which it was born, those very social formations were brought to birth by the ship. The terrestrial factory was premised upon the factories and manufactories in motion. Even the modes of surveillance of the prison rested upon the observation decks and crow's nest of the slave and the whale ship. In place of striation, stratification and sedimentation the sea offers different borders, civilities and sovereignties – ones which are enmeshed in the social memory of humans and other species.

In the first chapter we began to uncover the relations between species, nature and machines and introduced the centrality of animals and technology to our understanding of modernity. In the next chapter we focused specifically on Marx and his understanding of the machine and current Marxist attempts to rethink the question of nature as one of human domination. In Chapter 3 the 'Cetacean holocaust' was explored in both history and in fiction. In no way do we see them as comparable in order to reduce the horrors of the human holocaust but understanding the 'animal Treblinka' might help us understand the human capacity for destruction with the whaling 'industry' prefiguring the human extermination to come – comparative, but not equivalent. Chapter 4 began to illuminate the human encampments made amidst nature and ultimately extending into it and submerging. In Chapter 5 we tried to explicate what I call 'locomotive cultures': the prosthetic extension of the human hand which made mobility and the transformation of nature possible. In Chapter 6 we returned to questions of animal memory and nature and ways in which we can rethink our historiographical relationship to other species in terms of 'epistemic power'. In this chapter, we also returned to the question of human and animal labour and whether the 'metabolic rift' between humans and nature can be newly expressed in novel forms of solidarity between humans and other species. The concept of consciousness, sentience and memory might be central to new modes of wilding and conservation. Experimenting with both concepts and human labour and movement could delineate new modes of coexistence, cohabitation, integration and indeed separation.

Labour, reproduction and nature

Understanding the Cetacean holocaust entails the description of the human invasion of the habitations of other species. What we have not fully described are the ways in which segments of human labour, as appendages of machines and specifically the machines of the ships, were themselves compelled to subjection at the same time as other species. The dominion of the whale masters and iron masters over nature also entailed the subordination of subaltern 'semi-human' classes of humanity. Terry Eagleton in his analysis of Elizabeth Gaskell's *Sylvia's Lovers* has noted the sailors as the locus of proletarian class struggle against the masters but also against the 'repressive State-power symbolized by the press-gangs' (1976: 19). The novel itself is set in Scoresby's own whaling port of Whitby/Monkshaven and was the result of only a two-week sojourn that Gaskell spent there some years after the events it describes. Its context is the struggle between the land represented by Sylvia and the sea represented by the whaler Kinraid. Its uniqueness lies in its sense of isolation from other parts of humanity at the same time as the reassertion of a fundamental connection and cohabitation with nature.

The port itself is absolutely cut off from terrestrial civilisation and its only connections are oceanic into the vast distances of the north Atlantic and the polar ice. For Eagleton, the descriptions of the courage of the whalers, displays the fundamentally political status of their social class (1976: 19). The extension of the rule of the 'thanatocratic' state regime (Linebaugh 1991) into the whaling ships is both symbolic and physical. The isolated human encampments of the port and the ship are assailed by the military naval regimes of the press gangs wielding their power upon the decks of the 'wooden world' and inland into the moorland and vales of Yorkshire. Gaskell returns time and time again to one symbolic gesture and object: the knife in the hands of the whaler. At once this is used by the specksioneer Kinraid to carve the whale up and as a weapon against the state regime of the press gang. Eagleton notes this aspect of the whalers:

> …violence there is part of the fabric of ordinary social life, based as it is on a rhythm of tense hopes for the absent whalers, fierce excitement at their exploits, and anguish for the men lost. The knives used for the whales can, like Kinraid's, be turned against authority; the hunt to the death of the press-gang is part of a whaling society's daily existence.
>
> (Eagleton 1976: 19)

In 1779, the Whitby whaler Adamant returned from the Greenland Sea with eight whale carcases. Upon near entrance to the harbour it was boarded by a naval ship and all her hands were impressed before they could meet their families waiting for them on shore (Baines 2010: 94).

Stephen Schwartz has argued that *Sylvia's Lovers* is a symbolic articulation of sea and land and of two types of love; that for humans and that for God and 'life in the world' (1983/1984: 1). The land and the sea are both dichotomous and interrelated. But historiographically the sea has an entirely different status to that of the land. For Schwartz: 'Once the sea is recognized to be eternity, its destructive effect can be seen to be an illusion since human suffering and death are trivial to the soul in heaven', (1983/1984: 10). The destruction of both the whale and human life are simply seen as part of the dominion of God in which all is subject: whale, sailor and lover.

Not only is the sea 'eternal' and endlessly moving but warfare and destruction all come from the sea. The press gangs emerge from the the ocean depths as much as the carcases of whales and the seafaring dead. The littoral contact zone of the port is that space where the produce of the sea is unloaded from the decks onto oil and blubber yards, but also that place where dead sailors are unloaded from the ships. The littoral is also that place where the sailors are taken into the polar ice fields or into naval vessels and machines of war: indeed, literally the 'man of war'. Andrew Sanders, in his introduction to *Sylvia's Lovers,* has noted the central importance of Scoresby the younger to the genesis of the novel. Like Melville and Lovecraft, reading Scoresby's surveys and histories inspired the symbolic and fictional rendition of the polar darkness. Gaskell notes the unsavouriness of the trade and the genuinely amphibious and marine nature of the town and its inhabitants (1999: 1–2). But the people 'revelled in the odours of train-oil' (1999: 3). As Gaskell notes, whaling was central to the social identities and landscapes of these littoral places between sea and industrial civilisation:

> …for twenty miles inland there was no forgetting the sea, nor the sea-trade; refuse shell-fish, seaweed, the offal of the melting-houses, were the staple

> manure of the district; great ghastly whale-jaws, bleached bare and white, were the arches over the gate-posts to many a field or moorland stretch.
>
> (Gaskell 1999: 4–5)

Like the architectural bone assemblages of Thule and the polar north, the bones of the whales symbolised the continual presence of this destroyed species in the English landscape of pasture and yard.

The labouring whalers of the Greenland trade themselves underwent a transformative experience, leaving as young men and returning utterly destroyed by the experience of destruction and violence, if that is, they survived at all. As Gaskell writes: 'The whalers went out into the Greenland seas full of strong, hopeful men; but the whalers never returned as they sailed forth… Whose bones had been left to blacken on the gray and terrible icebergs? Who lay still until the sea should give up its dead?' (1999: 19). Unlike Scoresby's narratives, the labouring nautical proletariat come to life in Gaskell, indeed she is one of the few to archive and document the subjective experiences of this proletariat in formation albeit from the outside of those social relationships. The 'half-amphibious boys' of the port lead the struggles against the press gang and the state violence of the Napoleonic wars (1999: 28). As the press gang comes ashore, the 'Greek chorus' of women lead the assault against the state and the attempts to take their husbands and children. This is for Gaskell a great 'stormy multitude' (1999: 29). This multitude, as both chorus and as an abyssal and oceanic storm, literally submerges the agents of the state. But further, Gaskell describes their sheer *animality* in the face of social repression as the women are transformed into beasts, like the beasts of the sea and of the earth. Gaskell writes:

> Their wild, famished eyes were strained on faces they might not kiss, their cheeks were flushed to purple with anger or else livid with impotent craving for revenge. Some of them looked scarcely human; and yet an hour ago these lips, now tightly drawn back so as to show the teeth with the unconscious action of an enraged wild animal, had been soft and gracious with the smile of hope…
>
> (Gaskell 1999: 29)

Parallel to the social struggles on land were the struggles on the decks of the warships and the whalers and ultimately their struggles against bears and whales upon the great polar ice (1999: 100). Yet framing the narrative is God's dominion over, and utter destruction of, animals and of women and men. The graveyard is full of the whaling dead unloaded from ships and their bodies brought back to their women, others were lost in the Greenland seas, shipwrecked in the Baltic or drowned off the Icelandic coast (1999: 64). The ships and the machinic industry were not just about the submergence and elimination of nature, but the parallel destruction of the sentient human appendages who were compelled by subsistence to labour upon the decks of the wooden world.

Briton Cooper Busch's history of whale ship labour was one of the first to attempt an understanding of the production and reproduction of labour on the ships through a description of logs and journals (1994). Even more definitive was the publication of Martha Settle Putney's history from below of the African-American sailors in the trade – including whalers like Crispus Attucks and Paul Cuffee (1987: 2). The Atlantic interconnections not only connected the trades of the whale and the slave but also new iterations of proletarian civilisation across oceans (Rediker 1989). The 'official history' of the trade can in itself offer some clues to the 'subaltern' history of the nautical proletariat. Davis, Gallman and Gleiter's account of the institutional political economy of the whale trade offers outstanding insights into the accumulative power of the whale hunt (specifically based on the industry of New Bedford). Understanding the relations between productivity, the labour contract, the entrepreneurial managers and initiators of the trade and the market for its product (1997: 9), they explicate both the emergence and the collapse of the trade particularly around the economic productivity of labour and the 'product'.

More generally, the social production and reproduction of the body of the sailor and the sailor as labour offers some interesting insights into the nature of capital and its relation to the ocean. Paul Thompson and his collaborators, in their now classic, oral and archival history of ocean labour, have documented the complex social cultures and memories of the sailor. The ships themselves were situated in 'constellations' of economic, social and moral structures. Out of the sight of religious practice, if not God, there were tremendous issues about the social composition and stratification of the ships and the wider industry. These cultures were often highly suspicious of collective action and proletarian solidarity (Thompson et al. 1983: 5–6). The rise of steam-trawling supported the development of new kinds of social relations based on capital accumulation, wage labour and proletarianisation (1983: 18–19). But perhaps the most significant aspect of seafaring labour was the struggle around defending territorial rights to the sea (often masquerading as the conservation of fish). The 'communal property rights over the sea' would be part of the experience of generations of labour and contests between different cultures and nations would be part of the wider attempt to control nature (1983: 24). Ultimately a new 'herring capitalism' would create new kinds of spatial patterns as the ships followed the migrating shoals of herring (1983: 33). The struggles over governance and fishing would have an impact and restriction upon ancient rights of fishing – specifically as the idea of the sea as a 'commons' was replaced by international policy around fishing and conservation. The increasing proletarianisation of the mariner was itself perceived as a kind of serfdom, specifically as these labourers were in the bound cage of the ship for extended periods (1983: 51–52). There were complex systems of alliances and coalitions amongst different types of fishermen as new amalgamations, unions and provident societies were formed (1983: 82–83). 'Steam-capitalism' (1983: 110–111) would have an effect not just on labour, but upon the ecology of the ocean. As Thompson et al. note:

> The industry they created is indeed almost finished: and one can but gasp at the ruthless logic of a steam capitalism which had such power to suck men into its system, transform their work from a way of life into a form of bondage, break their family lives, degrade entire communities – and, ultimately, bring about its own destruction.
>
> (Thompson et al. 1983: 145)

Shepherding and capturing the whale, the proletariat was a class in oceanic motion. Its Atlantic interconnectivity created its hybrid and monstrous status as a class. Not only was it a 'many-headed hydra', subordinated as it were to the prosthetic machines of the ships, it was also part of a sentient resistance to the ship masters and to an emerging global capital which was enslaving human beings as well. These new 'transhumants' were in constant proletarian flux. Not only were they in combination with their weaponry and machines as objects of extension, they were also recombined as new social forces and products of multiple locations, identities and civilisations. They were indeed both a new evolutionary route for humans and a proto-revolutionary 'stormy multitude'. Their very animality and monstrousness made them more akin to the animals they were hunting than to the masters and financiers compelling their labour to utility on the polar ice. Degraded and subjected as they were, they were the basis for the new Atlantic working class. They were monstrous collages of ethnicity, identity and language with new coalescences and collaborations which would make and remake European, African and American civilisation. This was a diverse class-in-formation upon the sea (See Rediker 1989).

The concept of the monstrous multitude upon the ocean perhaps had its foremost prefiguration amongst the sailors of the Spithead and The Nore mutinies of 1797 – and specifically, in the last testament of the enigmatic Richard Parker. The mutinies are partly-documented but their significance, aside from their impact upon naval bureaucratic structures and conditions, remains largely unmarked (see Kemp 1970: 173–175 and Rodger 2004: 444–449). Perhaps the fullest account is that by Manwaring and Dobree (2004). They note that the horrors of service upon the naval ships was seen as just 'part of the order of nature' (2004: 9). The sailors were perceived as beasts by their officers, their animality paraded upon the deck (2004: 14). But in the mutinies of 1797 the sailors came together as a 'multitude' with a singular voice (2004: 16). The ships, belonging to the thanatocratic state paraded violence with the signs of the gallows bedecked around them or at least until their seizure by the sailors and their delegates and the flying of the 'bloody' red flag (2004: 50). The history of the Nore mutiny is especially instructive for understanding the bloody multitude of the rebels centred as it was around the *Sandwich*, '…an old corpse of a ship' (2004: 122). The delegates sailed round imperiously from ship to ship with flags, music and retainers with the foremost of the rebels, Richard Parker, in his enigmatic role of President of the Committee of the Delegates of the Fleet of the Nore (2004: 145). President of the great 'Floating Republic' (including the ship *Grampus* named for the whale and later to find its counterpart in Poe), Parker ultimately threatened that the rebels would be 'forced to repose in another country'. This was a hint that the fleet might be handed to the French or the Americans and there was even talk that they might make sail for the mythical utopian 'New Colony' (2004: 206–207). The concept of the 'multitude' would find its highest and most contradictory expression Parker's last testament before his hanging from the masts of the *Sandwich* (2004: 273–276).

In a completely different context the sociologist John Law notes something similar in his discussion of 'monstrous entities':

> Let me conclude by twisting Marx: I want to say that entities make history, but not in conditions of their own choosing. And, of course, I should add, dialectically, that entities are made by history too. Once we understand that entities and their relations are continuous; once we understand (as sociology does not) that they are heterogeneous; once we understand that the differences and distributions that are drawn between them could be otherwise; once we understand…that their histories and their fates vary widely; then we will come to appreciate that we are *all* monsters, outrageous and heterogeneous collages. And we will understand why it is that some monsters find it so easy that they scarcely look like monsters at all; why it is that some monsters are truly wretched, subjected to pain, deprived of all hope and dignity; and how it is that we might work towards a form of modest, multivocal organisation, where *all* could be reborn as hopeful monsters – as places where the necessary incompatibilities, inconsistencies and overlaps come gently and creatively together.
> (Law 1991: 18–19)

The idea of the 'hopeful monster' points to the prosthetic accrual to humans of technics but also to the 'organicisation' of mechanics and machines. Haraway's cyborgs, monstrous coalitions of flesh and technology become one of the proliferating models of a technical future. For Haraway, these new entities are: 'imploded entities, dense material semiotic "things" – articulated string figures of ontologically heterogeneous, historically-situated, materially rich, virally proliferating relatings of various sorts', (2012: 301). In actuality, these ideas of monstrousness and hybridity are the extension of the ideas of one book – Darwin's *The Origin of Species.* In his reworking of this book the geneticist Steve Jones has noted how much the notion of the *whale* was central to the epistemological revolution of evolutionary theory. He discusses hybridity in the context of Darwin's original remarks about the bear being 'almost like a whale' and the idea of the hybridised 'whale-bear' (1999: xvii–xviii). Darwin's vision of a bear in the endless transmutation of natural selection culminates in the whale as 'monstrous' (Jones 1997: xxvi). The transitional form of the whale is indeed itself simply a moment in that process of gradual monstrification (Jones 1997: 3). The 'descent' of the animal has to uncover its potential futures, re-elaborations and potential hybridities. As Jones notes: 'Take the aquatic bear for which Darwin suffered such mockery. Could it ever have made its way towards the sea? What does it take to become a whale, to live at the other end of the scale from a virus?' (1997: 17). The scale and bulk of the whale, its vast skeletal remains are indeed a world away in classificatory terms from the viral but their existence and genes reveal ancestries and descent. As Jones says of the bones of whales: 'Nobody will ever reconstruct the biography of whales from the fragments of their ancestors. Fortunately, whales – as much as viruses – are living fossils', (1997: 18–19). The 'fossilisation' and observation of the whale as a genetic organism reveals its morphological history. As Jones notes – the concept of a divinely ordered fixity of species was now dead (1997: xviii). Beginning with Darwin's understanding of gradual changes in the geology of the earth a new biological transmutation was being witnessed. For Jones 'If landscapes could be transfigured by slow change, so, surely, could flesh' (1997: xxv).

This transfiguration rested upon new understandings of the whale, their environment and their relation to other species. Darwin was specifically impressed, in his account of the voyage of the *Beagle* by the vastness and bulk of the cetacea and their meagre, as it seemed to him, food:

> If we suppose the case of the discovery of a skeleton of a Greenland whale in a fossil state, not a single cetaceous animal being known to exist, what naturalist would have ventured conjecture on the possibility of a carcass so gigantic being supported on the minute crustacea and mollusca living in the frozen seas of the extreme north?
>
> (Darwin 2009: 102n)

His sightings of whales would provide the factual basis for understanding whales in the process of transfiguration and mutation. Indeed, like the phantasms present for Marx at the birth of historical materialism (Hudson 2017) so the whale is present at the birth of evolutionary theory in *The Origin of Species*:

> We can also understand the apparent paradox, that the very same characters are analogical when one class or order is compared with another, but give true affinities when the members of the same class or order are compared with one another: thus the shape of the body and fin-like limbs are only analogical when whales are compared with fishes, being adaptations in both classes for swimming through the water; but the shape of the body and fin-like limbs serve as characters exhibiting true affinity between the several members of the whale family; for these cetaceans agree in so many characters, great and small, that we cannot doubt that they have inherited their general shape of body and structure of limbs from a common ancestor.
>
> (Darwin 2009: 866–867)

Rather than the bear, the whale resembles the modern otter: a mammalian swimmer just like its larger relative. In the transition from land to water some commentators have noted the importance of 'pelvic paddling and dorsoventral undulations of the tail' (Thewissen and Fish 1997: 482). The morphological evolution of whales has also led to significant localisation and genetic diversity amongst differing whale populations displaying their transmutation and the challenges faced when humans intervene in their ecologies (Hoelzel 1994). Indeed, human technological development can accelerate the processes of evolutionary monstrification or indeed potentially render extinct whole populations of whales like those around Nantucket in the eighteenth century. The ruptures and the punctuations in the evolutionary process have come to literal fruition in an Anthropocene epoch where both the flesh and the habitats of the whale are scored by human interventions. Yet it may be that the war against the animals is now being waged back against the 'human movement'. It might be that the sentient animals can help us elaborate new ways of thinking about proletarians and specifically the resisting proletarian animal itself.

The animal counter-republics

In an important essay from the early 1990s, Nancy Fraser proposed the development of the concept of the 'subaltern counterpublics' (1990). For Fraser: '…the bourgeois public was never *the* public. On the contrary, virtually contemporaneous with the bourgeois public there arose a host of competing counterpublics, including nationalist publics, popular peasant publics, elite women's publics, and working class publics', (1990: 61). The extension of the plebeian counterpublic contested the exclusions of bourgeois public spaces: to use a geological analogy again it contests the stratification discursively of domination and subordination (1990: 66). In obvious ways this argument rests upon Spivak's idea of the subaltern speaking (1988). Important as this is in terms of the reassertion of the discursive power of subaltern classes, sexualities and ethnic identities, can we transpose this concept to thinking about subaltern species? Specifically when we have already been told that the whale 'has no voice'? Indeed, surely proletarian subaltern subjectivity is about resistance to the discourses of mastery and power – the sentience of resistance which animals do not possess?

Dinesh Wadiwel, in his legal survey of human and animal sovereignty, has argued for an understanding of human-animal relations based upon warfare (2015: 3–4). The sovereignty of human power over entire territorial and biological regions of the planet (2015: 20) are enmeshed in the epistemic violence perpetrated against animals as an 'inferior entity' to human beings (2015: 34). Wadiwel and his collaborators have begun to development new understandings of the 'non–human' Anthropocene epoch (Human Animal Research Network Editorial Collective 2015). Yet this 'war against animals' is often perceived in 'social history' terms rather than in 'history from below'. When we ask questions about sentient animal resistance or whether animals have a voice, can we truly think of animals as a subaltern species comparable to subaltern proletarian class formations? Jason Hribal asserts that we can in his critical work on the animal proletariat and resistance (2010). Hribal argues for the convergence between the animal and the human proletariat and for the incorporation of animals into labour history. Entangled in the social struggles over the commons and enclosures, animals were subordinated to humans not as part of any kind of 'natural' evolution but because of the emergence of specific kinds of property and class formations. For Hribal, 'Animals do *not* "naturally" become private property, no more than humans "naturally" come to sell their labor. Rather, there is an *active* history here—one of expropriation, exploitation, and resistance', (2003: 436). Human dominion must be thereby expressed not in terms of species contestation but of social class, classification and subordination.

For Hribal the motive-power of pre-capitalist social formations was the animal. The animal was the provider of power for the development of agriculture, milling and transportation. This was, as Hribal argues, the primary reason why James Watt chose 'horse-power' as the metric by which steam-power was to be measured (2003: 443). The emergence of new property relationships intensified the need for 'animal-power' (2003: 443). The self-making of the industrial proletariat was accompanied by the rise in solidarity with other species:

> The publication of animal-concerned literature exploded forth in vast numbers. Indeed, at the exact moment as the formation of the working-class, we see the formation of the first collective, organized movement for the rights of animals in modern history. This was no coincidence.
>
> (Hribal 2003: 452)

The proletarianisation of both 'human and nonhuman alike' created literally new appetites for vegetarianism and for the construction of new networks of solidarity between humans and animals (2003: 453). Ultimately, the resistance of animals would be expressed not only in the social struggles to free whales from captivity but in animal resistance to the demands of its masters and against their use and imprisonment. Kasatka, a captive orca has shown resistance and aggression towards her keeper at Sea World. The agencies of captivity 'had to negotiate with their entertainers to extract labor and profit' (2006: 4).

The labour history from below of the plebeian animal proletariat is reviewed by Hribal in his discussion of William Hornaday, the collector of animal skeletons:

> While docked in Ireland outside of Belfast, he purchased several donkeys from a local resident, as he wanted their skeletons for an exhibition. On the road-side, the man killed the donkeys, and Hornaday began the process of skinning and gutting their bodies.
>
> (Hribal 2007: 101)

He was shortly assailed by a group of labourers and cottiers who demanded that he cease. As Hribal notes, this rested on discourses of solidarity between the cottiers and the donkeys:

> For the cottiers, donkeys could be property; they could be bought and sold; they could be worked. There were various rules and regulations regarding what, where, when, and how these animals lived and labored. There were punishments for misbehavior. There was sadism for pleasure. Yet, these creatures were not strictly thought of or treated as commodities, machines, or artifacts to dismemberment at will. There existed an intimate relationship between the cottager and the donkey. The donkeys had names. They had gender (he/she was not "it"). They had intelligence and reason. They had individual and unique personalities. They had spirituality. They had their own recognized customary rights. They participated in festivals. During May Day and Harvest Home, donkeys, along with cows, horses, and oxen, would be adorned with flowers and garlands. The individual animal's life had a value separate and outside of his or her production and reproduction. Jesus did once ride on the back of a jackass. The donkey was both an integrated and active member of the Irish habitus. This was a perspective from below.
>
> (Hribal 2007: 101)

Hribal notes that Hornaday had an entirely different epistemological relationship to the animals. His was one of division between the animal and human in which animals had no independent agency and was just in servitude to human will (2007: 101). His vivisection and analysis of the donkey skeleton was an observation and a history from 'above'. Hribal's methodology of animal history 'from below' concerns itself primarily, as he notes, with agency and includes the self-emancipation and resistance of the animal. But it also concerns itself with class – the relationships and solidarities built between animal and human labourers (2007: 102). Social history from above deprives animals of agency and subjectivity: 'The animals are not seen as agents. They are not active, as laborers, prisoners, or resistors. Rather, the animals are presented as static characters that have, over time, been used, displayed, and abused by humans. They emerge as objects — empty of any real substance', (2007: 102).

Animals who resisted the deployment of their labour by masters or tried to escape from captivity and bondage were often punished as recalcitrant. New spaces of punishment and captivity were created for these animals and often their bodies were scored and scarred by their masters to punish them, sometimes clipping wings or blinding the animals. Hribal argues that those animals that could not be broken and continued to resist were subject to capital punishment and executed. They were often hanged or decapitated publicly. Animals seeking autonomy were relentlessly pursued and eliminated: 'These were public executions: brutal in their methods, meaningful in their display, and purposeful in their intention. Societal violence towards animals became institutionalized', (2007: 104). Hribal argues that the captive animals were frequently placed into zoos and exhibitions, their dead bodies preserved and displayed: 'Humans became conditioned to see other creatures as products to exchange, specimens to collect, technology to utilize, or vermin to destroy', (2007: 104). But technological development itself was a response to animal labour and resistance with the proletarian animals replaced by the steam engine. For Hribal this was a mass 'replacement' of a whole class by machines (2007: 104). This was at the same time as human beings themselves, as we have seen, become the sentient appendages to the machines that they have created. For Hribal the intensity and the duration of animal labour with the acceleration of industry also meant the intensification of the methodologies of oppression. But it also increased the 'commonalities' in the struggles of the human and animal labouring class (2007: 105). The methods of physical punishment, surveillance, reward were responses to the resistance of the 'autonomous animals' (2007: 109).

The deployment of human and animal as sentient appendages of the machine had as its consequence modes of resistance and radical subjectivity towards the empire of capital. Understanding those sentient entities means understanding their modes of and resistance to capitalist exploitation. Anna Williams has argued for an understanding of the disciplinary regimes of the slaughterhouse which distort the animal's sentience in order to make the manufacture of their meat more efficient. She notes that the disciplinary regime of the slaughterhouse is one which subordinates both human and animal labour creating cross-species exploitation and solidarity (2004). These disciplinary regimes of observation and captivity might not be intended as methodologies of destruction but may be perceived as forms of conservation. Irus Braverman's empirical work on zoo management and administration has argued for an understanding of their zoological classificatory regimes as spectacular displays for human entertainment and education. But they are ones which subordinate other nonhuman species to the appetites of their observers (2012). Indeed, understanding animals 'from below' and modes of identity, difference and indistinction might lead us to profound new understandings of animal sentience and subjectivity (Calarco 2015).

Braverman's work on conservation and population biology has stressed an understanding of conservation in the classical terms of *in situ–ex situ* models. This explicates conservation in terms of captive and wild population observation and data capture and has implications for how those populations are managed. The identification, naming, recording and governance of species (2014: 26) is one which has an effect on the 'paradigm of viability' of entire species as populations are quantified and algorithmically expressed as part of new modes of governance (2014: 26). Yet some commentators have argued that living species have always been enmeshed in technology as 'industrializing organisms'. Indeed, new developments in biotechnology have their origin in the technological harnessing of nonhuman species. The war horse and the industrial horse have been central to the development of American capitalism and military acceleration (Greene 2004, 2008). Edward Russell has argued that we have to abandon the notion, initiated by Leo Marx (1964), of the 'machine in the garden' (and specifically the locomotive in the American wilderness). For Russell, an evolutionary history of technology must understand animals as technological organisms and as 'biological artifacts' in servitude to human expertise (2004: 1). The machine did not replace the animal or technology modify and transform nature, rather, humans have always used biological entities in the history of technics (2004: 1). Nicole Shukin has understood this relationship as 'animal capital'. Capital literally inscribes itself upon the bodies of animals rendered in servitude to human biopower (2009). The entanglement and enmeshment of technics with biology thereby concerns both humans and nonhumans. The prosthetic extension of human bodies has often been at the cost to the animal body but the animal itself has been entangled in technology for long durations in 'human' history.

Yet, as Jason Hribal has noted, the question of agency and class are central in understanding multi-species solidarities of the proletarian bodies and sentience of both human and nonhuman rather than seeing the human movement as ultimately in contestation with other species. Rather the master–slave dialectic of social domination contesting the working class of all species than human–animal warfare. The plebeian counter-republics of species in solidarity might prefigure new ways of thinking about observation and dominion.

The field, the forest and the Island

> The Fascist's passionate interest in animals, nature, and children is rooted in the lust to persecute. The significance of the hand negligently stroking a child's head, or an animal's back, is that it could just as easily destroy them…A creature is merely material for the master's bloody purposes. Thus the Führer takes the innocent into his service…
>
> Adorno and Horkheimer (1997: 253)

When Adorno and Horkheimer talk about 'the master's bloody purposes' they see the submission not just of one entity such as the child, or nature, or the animal but the entirety of subordination. The project of totalitarian domination is a will to dominate all. But solidarities and 'counter-republics' can counter this project of domination.

The concept of the plebeian counter-republic of human and nonhuman has significant implication for practices over nature. To archive and to voice the whale and indeed any animal or human, entails a counter-archival process to that of capital accumulation and the destructive utilisation of both human labour as sentient congealed capital and the animal body as tool to be used and as product to be consumed. It might also mean, like the 'republic' of Antarctica, that far from capturing animals and removing them into human spaces, human beings should remove their invasive presence from the territories of the animals. Continued observation and interaction may well continue but enmeshed in new forms of relations which abandon 'dominion' entirely. Indeed, the question of the future of nature is both urgent and in a certain sense redundant. The signals of the human neobiota in the geological record also signal that metabolic restoration of nature or the return of the wild is now largely impossible. The restoration or conservation of the remnants of archaic and ancient woodlands is often seen as a way of rethinking human intervention.

Conceptually this is a shift from the cultivated field to the wild forest, although no amount of radical restoration can dispel the human traces observable in the pollution of ring bark. Perhaps it is not for humans to re-imagine and prefigure the future of nature but to withdraw from their advance position in certain landscapes, territories and oceans. The counter-republic of animals (and not an animal kingdom) can only be valuable if it is allowed self-emancipation, self-definition and self-determination away from the human movement and its republics. The plebeian human-animal solidarities might point the way to new modes of interaction but also of separation and division between the human and other natures in order for other species to take their own evolutionary routes, far from being subjected to humans as 'industrializing organisms'. The metabolic rift then might not be restored but separated out as new human populations rethink their relationship to the land, to their encampments, and how they live upon and with the earth. It may indeed signal the re-emergence of social relations more like those of the prehistoric material cultures of cohabitation and coexistence with nature, as one species amongst many, and with complex relationships of ritual and care for those other species. The Mesolithic societies of Europe have often been studied as part of nature in an ecological manner rather than as social formations (Malone 2001: 26).

The submergence of humans back into nature away from the field into the forest hints at new forms of inhabitation and movement perhaps made possible, ironically, by technics and new forms of machines. Indeed, the human movement and species has to urgently rethink its relationship to nature and ecology. As Smol has noted:

> It is now clear that no part of our planet is safe from environmental degradation of some kind. The human "footprint" is now virtually everywhere, from the polar regions to the tallest mountains, to the tropical forests…Paleolimnology has shown that even isolated, seemingly pristine water bodies have been impacted by human activities. The rates may be very slow in some cases, perhaps imperceptible using conventional, short-term monitoring programs, but even remote ecosystems are changing due to human activities.
>
> (Smol 2008: 310)

Kohn, in his rethinking of a post-human anthropology has argued for understanding complex forest systems as 'thinking' entities much like sentient beings (2013) and it may be that new forms of habitation can be forged inside and with forests. But it also means abandoning the kinds of counter-enlightenment ideas that might sustain a new endarkenment. The whale oil that fuelled the lights of the locomotives must be surpassed but this does not mean that we do not have the capacity to have light.

Johan Rockström et al., in their new theorisation of Earth systems have argued for the definition of a set of 'planetary boundaries' which, if transgressed, threaten the future of humans as a species. The catastrophic consequences of these transgressions will signify a new non-linear, abrupt change in continental, planetary systems (2009: 1). The 'Holocene state' is threatened by these 'dangerous thresholds' (2009: 2). The acceleration of human driven 'species-extinction' leading to significant biodiversity loss is the first major extinction event to be human-designed (2009: 14–15). The spatial aspects of this process are noted by the authors specifically in terms of land-system boundaries and the fields of cropland (2009: 14) but it also implies another fundamental aspect of the catastrophe: not just the boundary and transgression as metaphors, and indeed as metaphors for a catastrophe centrally about humans and their future, but about a set of natural boundaries. This is that the dominion over nature, whether sanctioned by God or the Enlightenment project, may have to be surrendered by the human abdication of and abandonment of spaces occupied by other species, in particular, the ocean and its populations.

The project to understand what Adorno and Horkheimer called 'the indefatigable self-destructiveness of enlightenment' (1997: xi) is inseparable from the kinds of 'social freedom' made possible by the Enlightenment project (1997: xiii). Indeed, rather than focus on the 'conservation' of nature other opportunities might beckon: 'The task to be accomplished is not the conservation of the past, but the redemption of the hopes of the past' (1997: xv). The mastery of nature which the Enlightenment was compelled to be part of has to be rethought, entangled as it is in 'blind domination' and the 'submission of everything natural to the autocratic subject' (1997: xvi). For Adorno and Horkheimer: 'In the most general sense of progressive thought, the Enlightenment has always aimed at liberating men from fear and establishing their sovereignty. Yet the fully enlightened earth radiates disaster triumphant', (1997: 3). Yet that 'disaster triumphant' is also that moment when enlightenment displays its worst and its best intimations for humans and for nature. As Blanchot has said: 'It is dark disaster that brings the light' (1986: 7).

For Adorno and Horkheimer the Enlightenment project was one which aimed at the 'disenchantment' of the world, an end to myth and the reassertion of objective knowledge of humans and nature against unreason and 'fancy' (1997: 3). The entanglement of knowledge and power 'knows no obstacles' and knowledge is used by humans to dominate both nature and other humans (1997: 4). In this sense, for Adorno and Horkheimer 'Enlightenment is totalitarian' (1997: 6). The extension of human dominion ends up objectifying the 'spirits of men' in the opposite direction of an animism which 'spiritualised' an inert and docile object (1997: 28).

Yet for Adorno and Horkheimer, it is a specific part of nature that reveals the clue to domination – that social relation between the human and the animal (1997: 245–255). For them, the classificatory systems produced by human beings were deliberately constructed as methods of subjection and subordination to separate out the human from the animal. It created methods and modes of distinction (1997: 245). This rests upon the 'reason' that has been bestowed upon humans but not upon animals: 'Reason, mercilessly advancing, belongs to man. The animal, from which he draws his bloody conclusion, knows only irrational terror and the urge to make an escape from which he is cut off' (1997: 245). The cruelties perpetrated by humans upon their prey display the animal trying to flee end evade. Like the whale specifically, animality has 'no voice'.

> The want of reason has no words. Its possession, which dominates manifest history, is full of eloquence. The whole earth bears witness to the glory of man. Unreasoning creatures have encountered reason throughout the ages – in war and peace, in arena and slaughterhouse, from the lingering death-throes of the mammoth overpowered by a primitive tribe in the first planned assault down to the unrelenting exploitation of the animal kingdom in our own days. This visible process conceals the invisible from the executioners – existence denied the light of reason, animal existence itself.
>
> (Adorno and Horkheimer 1997: 245–246)

The animal is simply subject to the 'master's bloody purposes' (1997: 253). The tableau that Scoresby the elder displays upon and around the carcase of the great whale is itself the product of an ordained 'dominion'. But the dominion subjugates whole classes of people just as it does the whale. It prefigures the human holocausts that were yet to come.

Donald E. Pease, in his introduction to the new and completed text of C.L. R. James's philosophical reflection on Moby-Dick and Ahab, has noted that Ahab was considered by James to be the epitome of the 'totalitarian type' (2001: xiv). He was also for James an archetype of the 'Promethean man' bringing light into the world (2001: 9–10). For James, writing of Melville and of Ahab: 'Fire, power, the civilization of material progress, was a mighty creative force. But its creativity was mechanical. Mechanical is a word he will use many times. It is this which is destroying his life as a human being', (2001: 10). The 'enlightenment' of Ahab was also his destruction. Yet there is something else equally haunting about the figure of James at this moment of writing about Ahab and the *Pequod.* Interned as an alien on Ellis Island in 1952, he was desperate to be part of Melville's America. For Pease, in his commentary on James, Ellis island becomes an 'extraterritorial' mobile landscape, itself parallel to the 'floating culture' of the *Pequod* (2001: xxx). It is interstitial, temporary and oceanic. For James: 'The Island, like Melville's Pequod, is a miniature of all the nations of the world and all sections of society' (2001: 3). The refugee exile James, constantly moving from homeland to homeland, at once African, English and Colonial, is imprisoned in his own slave ship and a captive as he writes of the *Pequod* upon the sea. What is forwards for James upon this island and what is backwards? The traverse of the Atlantic abyss from landmass to landmass across vast swathes of ocean alerts us to what Engels called the 'non-historic': the ignored, the irrelevant, the subjected. James himself, imprisoned upon his island was discarded as disjecta membra but in a non-space, an extraterritorial location devoid of any future. Around him the flocks, herds and species gather and are hunted. The metabolic rift and the horrors of enlightenment at once destructive and liberating are schisms between classes and peoples as well as species. The mobile encampments of humans are just as precarious as the whale nations surrounding them.

The Puritan dominion that compelled Scoresby the elder to perform his bloody work also propelled his ships into the glacial landscapes of the north. Towards new continents and across the collisions of epochs. Like the slave ship, the whale ship is a chronotope for modernity (Gilroy 1993). The wooden world of its timbers congeal history, archives and the birth of modernity and enlightenment. The spirit of that world would indeed, as Max Weber had said, create a new 'polar night of icy darkness' in which the *Geist des Kapitalismus* would find its apotheosis (Weber 2001: xxxiii, xxv). The capital acquisitiveness of the slaver and the whaler would create new civilisations out of the destruction of others: of humans, habitats and species. It would ultimately create a metaphor of the 'iron cage' of rationality and of the 'polar night of icy darkness'. As Kalberg has said, this 'iron cage metaphor' was not of what was but of what 'might be' if these processes of rationality and acquisitiveness were unfettered (2001: 181). Yet the cage of the ship, of the bear, of the whalebone house are more than metaphors. The project of Enlightenment entailed both an accumulative knowledge about the earth, its species and its peoples, but it also provided the epistemological framework for the domination of nature and beings. The classificatory power of human knowledge ordered nature all the better to exploit it and ultimately bring both humans and nonhumans to the brink of extinction both by the suppression of individual species and global ecological catastrophe. The 'polar night of icy darkness' is both metaphorical and literal. The cage that James finds himself within, as he writes of the master and slave dialectics of Ahab and the *Pequod*, imprisoned in the liminal place between the old world of Europe and the new world of the Americas is symptomatic of the human and nonhuman predicament. His counter-archive of the dispossessed, of the whaler, of the whale, of the forgotten revolutionaries submerged by the passing epoch also marks a different old and new world. The epochal, glacial, continental transformation of the Holocene into the Anthropocene marks both disaster and possibility. It is a dialectic of luminescence, a dialectic of disaster and a disaster which brings the light. New worlds were delineated by James on his island surrounded by a world in motion, and encircled by ships and whales. Yet the *Pequod* still sails towards the dark polar night and Scoresby's polar bear still sleeps in exile in its iron cage.

Bibliography

Adorno, T.W. (1974) *Minima Moralia: Reflections from Damaged Life*. Trans. E.F.N. Jephcott. London: New Left Books.

Adorno, T.W. and Horkheimer, M. (1997) *Dialectic of Enlightenment*. Trans. J. Cumming. London: Verso.

Agamben, G. (1999) *Remnants of Auschwitz: The Witness and the Archive, Homo Sacer III*. Trans. D. Heller-Roazen. New York: Zone Books.

Agamben, G. (2004) *The Open: Man and Animal*. Trans. K. Attell. Stanford: Stanford University Press.

Agier, M. (2002) 'Between War and City: Towards an Urban Anthropology of Refugee Camps', *Ethnography*, 3(3): 317–366.

Agier, M. (2008) *On the Margins of the World: The Refugee Experience Today*. Cambridge: Polity.

Agier, M. (2010) 'Humanity as an Identity and Its Political Effects (A Note on Camps and Humanitarian Government)', *Humanity: An International Journal of Human Rights, Humanitarianism, and Development*, 1(1): 29–45.

Agier, M. (2011) 'From refuge the ghetto is born: Contemporary figures of heterotopias', in Hutchison, R. and Haynes, B.D. (Eds) *The Ghetto: Contemporary Global Issues and Controversies*. Boulder: Westview Press, 265–292.

Agier, M. and Bouchet-Saulnier, F. (2004) 'Humanitarian Spaces: Spaces of Exception', in Weissman, F. (Ed.), *The Shadow of Just Wars: Violence, Politics and Humanitarian Action*. Ithaca, New York: Cornell University Press, 297–313.

Alberts, P. (2011) 'Responsibility Towards Life in the Early Anthropocene', *Angelaki: Journal of the Theoretical Humanities*, 16(4): 5–17.

Ansel, W.D. (1978) *The Whaleboat: A Study of Design, Construction and Use from 1850–1970*. Mystic: Mystic Seaport Museum.

Applebaum, A. (2003) *Gulag: A History of the Soviet Camps*. London: Allen Lane.

Armitage, J. (1999) 'Ontological anarchy, the temporary autonomous zone, and the politics of cyberculture: A critique of Hakim Bey', *Angelaki: Journal of the Theoretical Humanities*, 4(2): 115–128.

Armstrong, R. (2014) 'After Machines: An Ecological Age of Space Exploration', *Journal of the British Interplanetary Society*, 67(07–09): 279–289.

Armstrong, R. (Ed.) (2016) *Star Ark: A Living, Self-Sustaining Spaceship*. Basel: Springer International Publishing.

Audouze, F. (2002) 'Leroi-Gourhan, a Philosopher of Technique and Evolution', *Journal of Archaeological Research*, 10(4): 277–306.

Austin, R.G. (1959) 'Virgil and the Wooden Horse', *Journal of Roman Studies*, 49(1/2): 16–25.

Aves, W.A.T. (2009) *The Railway Operating Division on the Western Front: The Royal Engineers in France and Belgium 1915–1919*. Donington: Shaun Tyas.

Aves, W.A.T. (2012) *Supporting the British Expeditionary Force: The Royal Engineer Railway Units in France 1939–40*. Donington: Shaun Tyas.

Bagge, S. (1987) *The Political Thought of The King's Mirror*. Odense: Odense University Press.

Bagge, S. (1994) 'Nature and Society in The King's Mirror', *Arkiv för nordisk filologi*, 109: 5–42.

Bailey, B. (1995) *George Hudson: The Rise and Fall of the Railway King*. Stroud: Alan Sutton.

Baines, S. (2010) *The Yorkshire Mary Rose: The Ship General Carleton of Whitby*. Pickering: Blackthorn Press.

Balasopoulos, A. (2006) '"Suffer a Sea Change": Spatial Crisis, Maritime Modernity, and the Politics of Utopia', *Cultural Critique*, 63: 122–156.

Barnard, A. (2000) *History and Theory in Anthropology*. Cambridge: Cambridge University Press.

Barnosky, A. (2014) *Dodging Extinction: Power, Food, Money, and the Future of Life on Earth*. Berkeley: University of California Press.

Barrow, T. (2001) *The Whaling Trade of North-East England 1750–1850*. Sunderland: Sunderland University Press.

Bekoff, M. (2003) 'Minding animals, minding earth: Science, nature, kinship, and heart', *Human Ecology Review*, 10(1): 56–76.

Bell, A. (2011) 'The Dialectic of Anthropocentrism', in Sanbonmatsu, J. (Ed.) *Critical Theory and Animal Liberation*. Washington, DC: Rowman & Littlefield Publishers, 163–175.

Bell, F. (1985) *At the Works: A Study of a Manufacturing Town*. London: Virago.

Benjamin, W. (1969) 'Theses on the Philosophy of History', in *Illuminations*. Trans. H. Zohn. New York: Schocken Books, 253–264.

Benton, T. (1989) 'Marxism and Natural Limits', *New Left Review*, 178: 51–81.

Berger, J. (1975) *A Seventh Man*. Harmondsworth: Penguin.

Berger, J. (1979) *Pig Earth*. London: Writers and Readers.

Berger, J. (1980) *About Looking*. London: Writers and Readers.

Berger, J. (1992) *And our faces, my heart, brief as photos*. London: Granta.

Berger, J. (1993) *Keeping a Rendezvous*. London: Granta.

Berger, J. (2001) *The Shape of a Pocket*. London: Bloomsbury.

Berryman, S. (2009) *The Mechanical Hypothesis in Ancient Greek Natural Philosophy*. Cambridge: Cambridge University Press.

Bey, H. (2003) *T.A.Z.: The Temporary Autonomous Zone, Ontological Anarchy, Poetic Terrorism*. New York: Autonomedia.

Billinghurst, J. (Ed.) (2000) *The Spirit of the Whale: Legend, History, Conservation*. Grantown-on-Spey: Colin Baxter.

Blanchot, M. (1986) *The Writing of the Disaster*. Trans. A. Smock. Lincoln: University of Nebraska Press.

Blaszak, M.W., Solomon, B., Gruber, J. and Guss, C. (2014) *Chicago: America's Railroad Capital, 1836 to Today*. Minneapolis: Voyageur Press.

Bloch, M. (1967a) 'Mediaeval Inventions', in *Land and Work in Mediaeval Europe: Selected Papers by Marc Bloch*. Trans. J.E. Anderson. London: Routledge, 169–185.

Bloch, M. (1967b) 'Technical change as a problem of collective psychology', in *Land and Work in Mediaeval Europe: Selected Papers by Marc Bloch*. Trans. J.E. Anderson. London: Routledge, 124–135.

Bloch, M. (1967c) 'The advent and triumph of the watermill', in *Land and Work in Mediaeval Europe: Selected Papers by Marc Bloch*. Trans. J.E. Anderson. London: Routledge, 136–168.

Bradley, S. (2015) *The Railways: Nation, Network and People*. London: Profile Books.

Braudel, F. (1975) *The Mediterranean and the Mediterranean World in the Age of Philip II, Vols One and Two*. Trans. S. Reynolds. London: Fontana Collins.

Braun, B. (2009) 'Towards a New Earth and a New Humanity: Nature, Ontology, Politics', in Castree, N. and Gregory, D. (Eds) *David Harvey: A Critical Reader*. Oxford: Blackwell, 191–222.

Braverman, I. (2012) *Zooland: The Institution of Captivity*. Stanford: Stanford University Press.

Braverman, I. (2014) 'Governing the Wild: Databases, Algorithms, and Population Models as Biopolitics', *Surveillance & Society*, 12(1): 15–37.

Bromley, A.G. (1986) 'Notes on the Antikythera Mechanism', *Centaurus*, 29(1): 5–27.

Bruce, J.D. (1913) 'Human Automata in Classical Tradition and Mediaeval Romance', *Modern Philology*, 10(4): 511–526.

Bruck J.N. (2013) 'Decades-long social memory in bottlenose dolphins', *Proceedings of the Royal Society B*, 280(1768): 1–6.

Brumbaugh, R.S. (1966) *Ancient Greek Gadgets and Machines*. New York: Crowell.

Bryant, J. (1998) '*Moby-Dick* as Revolution', in Levine, R.S. (Ed.) *The Cambridge Companion to Herman Melville*. Cambridge: Cambridge University Press, 65–90.

Burkett, P. (1996) 'On some common misconceptions about nature and Marx's critique of political economy', *Capitalism, Nature, Socialism*, 7(3): 57–80.

Burnett, D.G. (2007) *Trying Leviathan: The Nineteenth-Century New York Court Case That Put the Whale on Trial and Challenged the Order of Nature*. Princeton: Princeton University Press.

Burnett, D.G. (2012) *The Sounding of the Whale: Science and Cetaceans in the Twentieth Century*. Chicago: Chicago University Press.

Busch, B.C. (1994) *Whaling Will Never Do For Me: The American Whaleman in the Nineteenth Century*. Lexington: University of Kentucky Press.

Calarco, M. (2015) *Thinking Through Animals: Identity, Difference, Indistinction*. Stanford: Stanford University Press.

Casarino, C. (2002) *Modernity at Sea: Melville, Marx, Conrad in Crisis*. Minneapolis: University of Minnesota Press.

Castoriadis, C. (1984) *Crossroads in the Labyrinth*. Trans. K. Soper and M.H. Ryle. Brighton: The Harvester Press.

Castoriadis, C. (2003) 'Complexity, Magmas, History: The Example of the Medieval Town', in *The Rising Tide of Insignificancy*. Paris: Anonymous, 363–386.

Castree, N. (1996a) 'Birds, mice and geography: Marxisms and dialectics', *Transactions of the Institute of British Geographers*, 21(2): 342–362.

Castree, N. (1996b) 'Invisible Leviathan: Speculations on Marx, Spivak, and the Question of Value', *Rethinking Marxism*, 9(2): 45–78.

Castree, N. (1997) 'Nature, economy and the cultural politics of theory: the "war against the seals" in the Bering Sea, 1870–1911', *Geoforum*, 28(1): 1–20.

Castree, N. (1998) 'Scaling nature and the meta-theoretical imperative', *Capitalism, Nature, Socialism*, 9(3): 49–55.

Castree, N. (2000a) 'Between Mumford and Marx: Anatomising Imperial San Francisco', *Antipode*, 32(4): 494–499.

Castree, N. (2000b) 'Marxism and the Production of Nature', *Capital & Class*, 24(3): 5–36.

Castree, N. (2002) 'False antitheses? Marxism, nature and actor-networks', *Antipode*, 34(1): 111–146.

Castree, N. (2005) *Nature: The Adventures of a Concept*. London: Routledge.

Causley, C. (Ed.) (1978) *The Puffin Book of Salt-Sea Verse*. Harmondsworth: Puffin.

Chtcheglov, I. (1953) 'Formulary for a New Urbanism', *Bureau of Public Secrets*. Trans. K. Knabb. www.bopsecrets.org/SI/Chtcheglov.htm [Accessed 16 September 2015].

Clark, G. (1947) 'Whales as an Economic Factor in Prehistoric Europe', *Antiquity*, 21(82): 84–104.

Cleaver, H. (2000) *Reading Capital Politically*. Leeds: Anti/Theses.

Cole, T. (2003) *Holocaust City: The Making of the Jewish Ghetto*. London: Routledge.

Conneller, C. (2004) 'Becoming deer: corporeal transformations at Star Carr', *Archaeological Dialogues*, 11(1): 37–56.

Conneller, C. (2007) 'Inhabiting new landscapes: settlement and mobility in Britain after the last glacial maximum', *Oxford Journal of Archaeology*, 26(3): 215–237.

Conneller, C., Milner, N., Taylor, B. and Taylor, M. (2012) 'Substantial settlement in the European Early Mesolithic: new research at Star Carr', *Antiquity*, 86(334): 1004–1020.

Connor, R.C., Smolker, R.A. and Richards, A.F. (1992) 'Dolphin alliances and coalitions', in Harcourt, A.H. and de Waal, F.B.M. (Eds) *Coalitions and Alliances in Humans and Other Animals*. Oxford: Oxford University Press, 415–443.

Crutzen, P. and Stoermer, E. (2000) 'The "Anthropocene"', *IGBP Newsletter*, 41: 17–18.

Dale, R. (1994) *Early Railways*. London: The British Library.

Darwin, C. (2009) *The Origin of Species and The Voyage of the Beagle*. London: Vintage.

Davis, L.E., Gallman, R.E. and Gleiter, K. (1997) *In Pursuit of Leviathan: Technology, Institutions, Productivity, and Profits in American Whaling, 1816–1906*. Chicago: Chicago University Press.

Derrida, J. (1976) *Of Grammatology*. Trans. G. Chakravorty Spivak. Baltimore: John Hopkins University Press.

Derrida, J. (1995) *Points: Interview, 1974–1994*. Trans. P. Kamuf. Ed. E. Weber. Stanford: Stanford University Press.

Derrida, J. (1996) *Archive Fever: A Freudian Impression*. Trans. E. Prenowitz. Chicago University of Chicago Press.

Donath, J. (2012) *The Social Machine: Designs for Living Online*. Cambridge: MIT Press.

Drachmann, A.G. (1968) *The Mechanical Technology of Greek and Roman Antiquity: A Study of the Literary Sources*. Copenhagen: Munksgaard.

DuBois, P. (2003) *Slaves and Other Objects*. Chicago: University of Chicago Press.

Durkheim, E. (1915) *The Elementary Forms of the Religious Life*. Trans. J. Ward Swain. London: Allen and Unwin.

Durkheim, E. (2001) *The Elementary Forms of the Religious Life*. Trans. C. Cosman. Oxford: Oxford University Press.

Dyball, R. and Newell, B. (2015) *Understanding Human Ecology: A systems approach to sustainability*. London: Routledge.

Eagleton, T. (1976) '"Sylvia's Lovers" and Legality', *Essays in Criticism*, XXVI(1): 17–27.

Edkins, J. (2000) 'Sovereign Power, Zones of Indistinction, and the Camp', *Alternatives: Global, Local, Political*, 25(1): 3–25.

Elliott, B.J. and Milner, N. (2010) 'Making a point: a critical review of the barbed point manufacturing process practiced at Star Carr', *Proceedings of the Prehistoric Society*, 76: 75–94. DOI: https://doi.org/10.1017/S0079497X00000451

Elster, J. (1983) *Explaining Technical Change: A Case Study in the Philosophy of Science*. Cambridge: Cambridge University Press.

Engels, F. (1934) *Anti-Dühring: Herr Eugen Dühring's Revolution in Science*. Trans. E. Burns. Ed. by C.P. Dutt. London: Lawrence & Wishart.

Engels, F. (1941) *Dialectics of Nature*. Trans. C. Dutt. London: Lawrence and Wishart.

Engels, F. (1954) *Dialectics of Nature*. Rev. Trans. C. Dutt. Moscow: Foreign Languages Publishing House.

Engels, F. (1969) *The Condition of the Working Class in England*. Trans. The Institute of Marxism–Leninism. London: Panther.

Esposito, R. (2011) 'Politics and Human Nature'. Trans. L. Chiesa, *Angelaki: Journal of the Theoretical Humanities*, 16(3): 77–84.

Fabian, J. (2001) *Anthropology with an Attitude: Critical Essays*. Stanford: Stanford University Press.

Fischer, E. (1963) *The Necessity of Art: A Marxist Approach*. Trans. A. Bostock. Harmondsworth: Penguin.

Foster, J.B. (1999) 'Marx's Theory of Metabolic Rift', *American Journal of Sociology*, 105(2): 366–405.

Foster, J.B. (2000) *Marx's Ecology: Materialism and Nature*. New York: Monthly Review Press.

Foster, J.B. and Burkett, P. (2016) *Marx and the Earth: An Anti-Critique*. Boston: Brill.

Godard. R. (1991) 'Long-term memory of individual neighbours in a migratory songbird', *Nature*, 350: 228–229.

Fyfe, P. (2015) *By Accident or Design: Writing the Victorian Metropolis*. Oxford: Oxford University Press.

Gaskell, E. (1999) *Sylvia's Lovers*. Oxford: Oxford University Press.

Gaskin, R.T. (1909) *The Old Seaport of Whitby*. Whitby: Forth and Son.

Geras, N. (1983) *Marx and Human Nature: Refutation of a Legend*. London: Verso.

Gerlach, C. (2000) 'The Wannsee Conference, the fate of German Jews, and Hitler's decision in principle to exterminate all European Jews', in Bartov, O. (Ed.) *The Holocaust: Origins, Implementation, Aftermath*. London: Routledge, 106–161.

Gigliotti, S. (2006) 'Cattle Car Complexes: A Correspondence with Historical Captivity and Post-Holocaust Witnesses', *Holocaust and Genocide Studies*, 20(2): 256–277.

Gigliotti, S. (2009) *The Train Journey: Transit, Captivity, and Witnessing in the Holocaust*. New York: Berghahn Books.

Gilbert, M. (1986) *The Holocaust: The Jewish Tragedy*. London: HarperCollins.

Gillespie, K. and Collard, R-C. (Eds) (2015) *Critical Animal Geographies: Politics, Intersections and Hierarchies in a Multispecies World*. London: Routledge.

Gilroy, P. (1993) *The Black Atlantic: Modernity and Double Consciousness*. London: Verso.

Gimpel, J. (1983) *The Cathedral Builders*. Trans. T. Waugh. Salisbury: Michael Russell.

Gimpel, J. (1992) *The Medieval Machine: The Industrial Revolution of the Middle Ages*. London: Pimlico.

Ginzburg, C. (1990) *Myths, Emblems, Clues*. Trans. J. and A.C. Tedeschi. London: Hutchinson Radius.

Gittins, S. (2010) *The Great Western Railway in the First World War*. Stroud: The History Press.

Godard. R. (1991) 'Long-term memory of individual neighbours in a migratory songbird', *Nature*, 350: 228–229.

Goode, C. (2015) *The Forest and the Field: Changing theatre in a changing world*. London: Oberon Books.

Gracq, J. (1986) *Le Rivage des Syrtes* [The Opposing Shore]. Trans. R. Howard. New York: Columbia University Press.

Greene, A.N. (2004) 'War Horses: Equine Technology in the American Civil War', in Schrepfer, S. and Scranton, P. (Eds) *Industrializing Organisms: Introducing Evolutionary History*. New York: Routledge, 143–166.

Greene, A.N. (2008) *Horses at Work: Harnessing Power in Industrial America*. Harvard: Harvard University Press.

Grossman, V. (2011) 'The Hell of Treblinka'. Trans. R. Chandler, in Rajchman, C., *Treblinka: A Survivor's Memory 1942–1943*. London: Maclehose Press, 113–181.

Haff, P.K. (2014a) 'Humans and Technology in the Anthropocene: Six Rules', *The Anthropocene Review*, 1(2): 126–130.

Haff, P.K. (2014b) 'Technology as a geological phenomenon: implications for human well-being', *Geological Society Special Publication*, 395:1: 301–309.

Hannam, K., Sheller, M. and Urry, J. (2006) 'Editorial: Mobilities, Immobilities and Moorings', *Mobilities*, 1(1): 1–22.

Haraway, D. (2008) *When Species Meet*. Minneapolis: University of Minnesota Press.

Haraway, D. (2012) 'Awash in Urine: DES and Premarin® in Multispecies Response-ability', *WSQ: Women's Studies Quarterly*, 40(1): 301–316.

Haraway, D. (2015) 'Anthropocene, Capitalocene, Plantationocene, Chthulucene: Making Kin', *Environmental Humanities*, 6(1): 159–165.

Hardt, M. and Negri, A. (2001) *Empire*. Cambridge: Harvard University Press.

Harkness, R. (2014) 'On Stone Houses and the Co-Creation of Worlds and Selves', in Souvatzi, S.G. and Hadji, A. (Eds) *Space and Time in Mediterranean Prehistory*. London: Routledge, 64–83.

Harman, G. (2010) 'Technology, objects and things in Heidegger', *Cambridge Journal of Economics*, 34(1): 17–25.

Harman, G. (2012) *Weird Realism: Lovecraft and Philosophy*. Alresford: Zero Books.

Hartigan, J. (2015) *Aesop's Anthropology: A Multispecies Approach*. Minneapolis: University of Minnesota Press.

Hay, D., Linebaugh, P., Rule, J.G., Thompson, E.P. and Winslow, C. (1976) *Albion's Fatal Tree: Crime and Society in Eighteenth-Century England*. New York: Pantheon.

Heidegger, M. [1926](1962) *Being and Time* ['Sein und Zeit']. Trans. J. Macquarrie and E. Robinson. Oxford: Basil Blackwell.

Hoare, P. (2009) *Leviathan or, The Whale*. London: Fourth Estate.

Hoelzel, E.R. (1994) 'Genetics and ecology of whales and dolphins', *Annual Review of Ecology and Systematics*, 25: 377–399.

Hribal, J. (2003) '"Animals are Part of the Working Class": A Challenge to Labor History', *Labor History*, 44(4): 435–453.

Hribal, J. (2006) 'When Animals Resist Their Exploitation: Kasatka, the Sea World Orca', *Counterpunch*, 14 December, 1–4.

Hribal, J. (2007) 'Animals, Agency, and Class: Writing the History of Animals from Below', *Human Ecology Review*, 14(1): 101–112.

Hribal, J. (2010) *Fear of the Animal Planet: The Hidden History of Animal Resistance.* New York: Counterpunch.

Hudson, M. (2000) 'The Clerk of the Foresters Records: John Berger, the Dead, and the Writing of History', *Rethinking History: The Journal of Theory and Practice*, 4(3): 261–279.

Hudson, M. (2002) 'On the dead of world history', *Race and Class*, 43(4): 26–33.

Hudson, M. (2015a) 'The Anthropocene and a critical theory of machines', *Heathwood Institute and Press*, September 2015, www.heathwoodpress.com [Accessed 28 September 2015].

Hudson, M. (2015b) 'The Empire of Neolithia', *Heathwood Institute and Press*, November 2015, www.heathwoodpress.com/the-empire-of-neolithia-martyn-hudson/ [Accessed 15 October 2016].

Hudson, M. (2016a) *Centaurs, Rioting in Thessaly: Memory and the Classical World.* New York: Punctum.

Hudson, M. (2016b) *The Slave Ship, Memory and the Origin of Modernity.* London: Routledge.

Hudson, M. (2017) *Ghosts, Landscapes and Social Memory.* London: Routledge.

Human Animal Research Network Editorial Collective (Eds) (2015) *Animals in the Anthropocene: Critical Perspectives on Non-human Futures.* Sydney: Sydney University Press.

Ingold, T. (2002) 'Between evolution and history: Biology, culture and the myth of human origins', in Wheeler, N. and Ziman, J. (Eds) *The Evolution of Cultural Entities.* Oxford: Oxford University Press, 43–66.

Ingold, T. (2013) *Making: Anthropology, Archaeology, Art and Architecture.* London: Routledge.

James, C.L.R. (2001) *Mariners, Renegades and Castaways: The Story of Herman Melville and the World We Live In, The Complete Text.* Hanover: University Press of New England.

Jamieson, D. and Regan, T. (1984) 'Whales are not cetacean resources', in Fox, M.W. and Mickley, L.D. (Eds) *Advances in animal welfare science 1984/85.* Washington, D.C.: The Humane Society of the United States, 101–111.

Jones, H. (1973) *Steam Engines: An International History.* London: Ernest Benn.

Jones, J.S. (1999) *Almost Like a Whale:* The Origin of Species *Updated.* New York: Doubleday.

Jünger, E. (1970) *On the Marble Cliffs.* Trans. S. Hood. Harmondsworth: Penguin.

Jünger, E. (2004) *Storm of Steel.* Trans. M. Hoffman. Harmondsworth: Penguin.

Kalberg, S. (2001) 'The Modern World as a Monolithic Iron Cage? Utilizing Max Weber to Define the Internal Dynamics of the American Political Culture Today', *Max Weber Studies*, 1(2): 178–195.

Kalechofsky, R. (2003) *Animal Suffering and the Holocaust: The Problem with Comparisons.* Boston: Micah Publications.

Kemp, P. (1970) *The British sailor: A social history of the lower deck.* London: J.M. Dent and Sons.

Kirk, S. (2005) *Philip Webb: Pioneer of the Arts and Crafts Movement.* Chichester: Wiley-Academy.

Koetsier, T. (2001) 'On the prehistory of programmable machines: musical automata, looms, calculators', *Mechanism and Machine Theory*, 36: 589–603.

Kohn, E. (2013) *How Forests Think: Toward an Anthropology Beyond the Human.* Oakland: University of California Press.

Kovel, J. (2002) *The Enemy of Nature: The End of Capitalism or the End of the World?* London: Zed Books.
Law, J. (Ed.) (1991) *A Sociology of Monsters: Essays on Power, Technology and Domination*. London: Routledge.
Leakey, R. (1994) *The Origin of Humankind*. London: Weidenfeld & Nicolson.
Lecourt, D. (1977) *Proletarian Science? The Case of Lysenko*. Trans. B. Brewster. London: New Left Books.
Lee, P.Y. (2008) 'Introduction: Housing Slaughter', in Lee, P.Y. (Ed.) *Meat, Modernity, and the Rise of the Slaughterhouse*. Lebanon: University of New Hampshire Press, 1–9.
Lee, R. (1990) 'Animal Liberation but not too much', *Arkangel*, 2, Spring 1990.
Lemmens, P. (2011) '"This system does not produce pleasure anymore": An Interview with Bernard Stiegler', *Krisis: Journal of Contemporary Philosophy*, 1: 33–41.
Leroi-Gourhan, A. (1993) *Gesture and Speech*. Trans. A. Bostock Berger. Cambridge: MIT Press.
Levi, P. (1979) *If this is a Man/The Truce*. Trans. S. Woolf. Harmondsworth: Penguin.
Lewis, M.J.T. (1970) *Early Wooden Railways*. London: Routledge & Kegan Paul.
Lewis-Williams, D. (2002) *The Mind in the Cave: Consciousness and the Origins of Art*. London: Thames & Hudson.
Lewis-Williams, D. and Pearce, D. (2005) *Inside the Neolithic Mind: Consciousness, Cosmos and the Realm of the Gods*. London: Thames & Hudson.
Lilly, J.C. (1967) *The Mind of the Dolphin: A Nonhuman Intelligence*. New York: Doubleday.
Linebaugh, P. (1982) 'All the Atlantic Mountains Shook', *Labour/Le Travailleur*, 10: 87–121.
Linebaugh, P. (1991) *The London Hanged: Crime and Civil Society in the Eighteenth Century*. Cambridge: Cambridge University Press.
Linzey, A. (1991) 'The Servant Species: Humanity as Priesthood', *Between the Species: A Journal of Ethics*, 6(3): 109–120.
Linzey, A. (1993) 'Liberation Theology and the Oppression of Animals', *Scottish Journal of Theology*, 46(4): 507–525.
Linzey, A. (1994) *Animal Theology*. London: SCM Press.
Linzey, A. (2001) 'Against biodiversity', *The Animals' Agenda*, March/April 2001, 21.
Linzey, A. (2007) *Creatures of the Same God. Explorations in Animal Theology*. Winchester: Winchester University Press.
Linzey, A. and Cohn-Sherbok, D. (1997) *After Noah: Animals and the Liberation of Theology*. London: Mowbray.
Little, A., Elliott, B., Conneller, C., Pomstra, D., Evans, A.A., Fitton, L.C., Holland, A., Davis, R., Kershaw, R., O'Connor, S., O'Connor, T., Sparrow, T., Wilson, A.S., Jordan, P., Collins, M.J., Colonese, A.C., Craig, O.E., Knight, R., Lucquin, A.J.A., Taylor, B., and Milner, N. (2016). 'Technological Analysis of the World's Earliest Shamanic Costume: A Multi-Scalar, Experimental Study of a Red Deer Headdress from the Early Holocene Site of Star Carr, North Yorkshire, UK', *PLoS ONE*, 11(4), e0152136, https://doi.org/10.1371/journal.pone.0152136 [Accessed 10 October 2016].
Lovecraft, H.P. (2007) 'At the Mountains of Madness', in *The Whisperer in Darkness: Collected Stories, Volume One*. Ware: Wordsworth, 277–370.
Lovegrove, K. (2004) *Railway: Identity, Design and Culture*. London: Laurence King.

Low, P. (2012) 'The Cambridge Declaration on Consciousness', 7 July 2012, at the *Francis Crick Memorial Conference on Consciousness in Human and non-Human Animals*, Churchill College, University of Cambridge. https://web.archive.org/web/20131109230457/http://fcmconference.org/img/CambridgeDeclarationOnConsciousness.pdf [Accessed 24 August 2016].

Lowell, R. (1996) 'The Quaker Graveyard in Nantucket', in Ferguson, M., Salter, M.J. and Stallworthy, J. (Eds) *The Norton Anthology of Poetry. 4th Edition*, 1485–1488.

Lozowick, Y. (1991) '"Judenspediteur": A Deportation Train', *Holocaust and Genocide Studies*, 6(3): 283–292.

Lucas, G. and Edwald, Á. (2015) 'Capitalism and Mobility in the North Atlantic', in Leone, M.P. and Knauf, J.E. (Eds) *Historical Archaeologies of Capitalism, 2nd Edition*. Heidelberg: Springer, 227–248.

MacKenzie, D.A. (1984) 'Marx and the Machine', *Technology and Culture*, 25(3): 473–502.

MacKenzie, D.A. (1998) *Knowing Machines: Essays on Technical Change*. Cambridge: MIT Press.

Malone, C. (2001) *Neolithic Britain and Ireland*. Stroud: Tempus.

Mann, J., Connor, R.C., Tyack, P.L. and Whitehead, H. (2000) *Cetacean societies: field studies of dolphins and whales*. Chicago: University of Chicago Press.

Manwaring, G.E. and Dobree, B. (2004) *The Floating Republic: An Account of the Mutinies at Spithead and The Nore in 1797*. Barnsley: Pen and Sword Books.

Marcus, S.P. (1974) *Engels, Manchester and the Working Class*. New York: Random House.

Marias, J. (2001) *Dark Back of Time*. Trans. E. Allen. Harmondsworth: Penguin.

Marino, L. (2002) 'Convergence of complex cognitive abilities in cetaceans and primates', *Brain Behavior and Evolution*, 59(1–2): 21–32.

Marino, L. and Frohoff, T. (2011) 'Towards a new paradigm of non-captive research on cetacean cognition', *PloS One*, 6(9), e24121.

Marx, K. (1964) *Pre-Capitalist Economic Formations*. Trans. J. Cohen. London: Lawrence & Wishart.

Marx, K. (1973) *Grundrisse: Foundations of the Critique of Political Economy (Rough Draft)*. Trans. M. Nicolaus. Harmondsworth: Penguin/New Left Books.

Marx, K. (1975) *Early Writings*. Trans. R. Livingstone and G. Benton. New York: Vintage.

Marx, K. (1976) *Capital: A Critical Analysis of Capitalist Production, Vol. One*. Trans. S. Moore and E. Aveling. Ed. by F. Engels and D. Torr. London: George Allen and Unwin.

Marx, K. and Engels, F. (1972) *Marx-Engels Reader*. Ed. and Trans. by R.C. Tucker. New York: W.W. Norton & Co.

Marx, K. and Engels, F. (1975) *Marx-Engels: Collected Works*. New York: International Publishers.

Marx, K. and Engels, F. (1974) *The German Ideology, Part One*. Ed. C.J. Arthur, London: Lawrence & Wishart.

Marx, K. and Engels, F. (1992) *The Communist Manifesto*. Trans. S. Moore. Ed. and Rev. by D. McLellan. Oxford: Oxford University Press.

Marx, L. (1964) *The Machine in the Garden: Technology and the Pastoral Ideal in America*. New York: Oxford University Press.

Matus, J.L. (2009) *Shock, Memory and the Unconscious in Victorian Fiction*. Cambridge: Cambridge University Press.

McComb, K., Moss, C., Durant, S., Baker, L. and Sayialel, S. (2001) 'Matriarchs act as repositories of social knowledge in African elephants', *Science*, 292: 491–494.

McKinnon, A.M. (2014) 'Elementary Forms of the Metaphorical Life: Tropes at Work in Durkheim's Theory of the Religious', *Journal of Classical Sociology*, 14(2): 203–221.

McLellan, D. (1971) *Marx's Grundrisse.* London: Macmillan.

Melville, H. (1992) *Moby-Dick or, The Whale.* Harmondsworth: Penguin.

Milner, N., Taylor, B., Conneller, C. and Schadla-Hall, R.T. (2013) *Star Carr: Life in Britain after the Ice Age.* London: Council for British Archaeology.

Milner, N., Bamforth, M., Beale, G., Carty, J.C., Chatzipanagis, K., Croft, S., Conneller, C., Elliot, B., Fitton, L.C., Knight, B., Kroger, R., Little, A., Needham, A., Robson, H.K., Rowley, C.C.A., and Taylor, B. (2016) 'A unique engraved shale pendant from the site of Star Carr: the oldest Mesolithic art in Britain', *Internet Archaeology*, 40, http://dx.doi.org/10.11141/ia.40.8.

Mithen, S. (1998) *The Prehistory of the Mind: A Search for the Origins of Art, Religion and Science.* London: Phoenix.

Morris, S.P. (1992) *Daidalos and the Origins of Greek Art.* Princeton: Princeton University Press.

Mulville, J. (2002) 'The role of cetacea in prehistoric and historic Atlantic Scotland', *International Journal of Osteoarchaeology*, 12(1): 34–48.

Museum of Modern Art [MOMA] (1994) *Machine Art, March 6 to April 30, 1934.* New York: Museum of Modern Art.

Nansen, F. (1911a) *In Northern Mists: Arctic Exploration in Early Times, Volume One.* Trans. A.G. Chater. London: William Heinemann.

Nansen, F. (1911b) *In Northern Mists: Arctic Exploration in Early Times, Volume Two.* Trans. A.G. Chater. London: William Heinemann.

Negri, A. (1991) *Marx Beyond Marx: Lessons on the Grundrisse.* Trans. H. Cleaver, M. Ryan and M. Viano. Ed. by J. Fleming. New York: Autonomedia/Pluto.

Neves, K. (2010) 'Cashing in on Cetourism: A Critical Engagement with Dominant E-NGO Discourses on Whaling, Cetacean Conservation, and Whale Watching', *Antipode on Capitalism and Conservation: Special Issue*, 42(3): 719–741.

Neves-Graca, K. (2004) 'Revisiting the Tragedy of the Commons: Whale Watching in the Azores and its Ecological Dilemmas', *Human Organisation*, 63(3): 289–300.

Nibert, D. (2014) 'Animals, Immigrants, and Profits: Slaughterhouses and the Political Economy of Oppression', in Sorenson, J. (Ed.) (2014) *Critical Animal Studies: Thinking the Unthinkable.* Toronto: Canadian Scholars' Press, 3–17.

Nocella, A.J., Sorenson, J., Socha, K. and Matsuoka, A. (Eds) (2014) *Defining Critical Animal Studies: An Intersectional Social Justice Approach for Liberation.* New York: Peter Lang.

Nyman, J. and Schuurman, N. (Eds) (2016) *Affect, Space and Animals.* London: Routledge.

O'Connor, J. (1989) 'Capitalism, Nature, Socialism: A Theoretical Introduction', *Capitalism, Nature, Socialism*, 1(1): 11–38.

Olivius, E. (2013) 'Gender Equality and Neo-Liberal Governmentality in Refugee Camps', *St Antony's International Review*, 9(1): 53–69.

Olsen, B., Shanks, M., Webmoor T. and Witmore, C. (2012) *Archaeology: The Discipline of Things.* Berkeley: University of California Press.

Pacyga, D.A. (2008) 'Chicago: Slaughterhouse to the World', in Lee, P.Y. (Ed.) *Meat, Modernity, and the Rise of the Slaughterhouse.* Lebanon: University of New Hampshire Press, 153–166.

Patterson, C. (2002) *Eternal Treblinka: Our Treatment of Animals and the Holocaust.* New York: Lantern Books.

Philbrick, N. (2000) *In the Heart of the Sea: The Tragedy of the Whaleship* Essex. London: HarperCollins.

Pick, D. (1996) *War Machine: The Rationalization of Slaughter in the Modern Age.* New Haven: Yale University Press.

Poe, E.A. (1998) *The Narrative of Arthur Gordon Pym and Related Tales.* Oxford: Oxford University Press.

Postone, M. (1993) *Time, Labor and Social Domination: A Reinterpretation of Marx's Critical Theory.* New York: Cambridge University Press.

Purdy, J. (2015) *After Nature: A Politics for the Anthropocene.* Cambridge: Harvard University Press.

Putney, M.S. (1987) *Black Sailors: Afro-American Merchant Seamen and Whalemen Prior to the Civil War.* New York: Greenwood Press.

Rackham, O. (2015) *Woodlands.* London: William Collins.

Ramadan, A. (2009) 'Destroying Nahr el-Bared: Sovereignty and urbicide in the space of exception', *Political Geography,* 28(3): 53–163.

Ramadan, A. (2010) 'In the Ruins of Nahr al-Barid: Understanding the meaning of the camp', *Journal of Palestine Studies,* 40(1): 49–62.

Ramadan, A. (2013) 'Spatialising the refugee camp', *Transactions of the Institute of British Geographers,* 38(1): 65–77.

Raunig, G. (2010) *A Thousand Machines: A Concise Philosophy of the Machine as Social Movement.* Trans. A. Derieg. Los Angeles: Semiotext(e).

Redhead, S. (2004) *Paul Virilio: Theorist for an Accelerated Culture.* Toronto: University of Toronto Press.

Rediker, M. (1989) *Between the Devil and the Deep Blue Sea: Merchant Seamen, Pirates and the Anglo-American Maritime World, 1700–1750.* Cambridge: Cambridge University Press.

Rediker, M. (2008) 'History from below the water line: Sharks and the Atlantic slave trade', *Atlantic Studies,* 5(2): 285–297.

Redman, N. (2004) *Whales' Bones of the British Isles.* Redman: Redman Publishing.

Ribeiro, A. (2002) *Dress in Eighteenth Century Europe 1715–1789.* New Haven: Yale University Press.

Richerson, P.J. and Boyd, R. (1998) 'The Evolution of Human Ultra-sociality', in Eibl-Eibesfeldt, I. and Salter, F. (Eds) *Indoctrinability, Ideology, and Warfare: Evolutionary Perspectives.* New York: Berghahn Books, 71–95.

Rockström, J., Steffen, W., Noone, K., Persson, Å., Chapin, F.S., Lambin, E., Lenton, T.M., Scheffer, M., Folke, C., Schellnhuber, Nykvist, H.B., De Wit, C.A., Hughes, T., van der Leeuw, S., Rodhe, Sörlin, H.S., Snyder, P.K., Costanza, R., Svedin, U., Falkenmark, M., Karlberg, L., Corell, R.W., Fabry, V.J., Hansen, J., Walker, B., Liverman, D., Richardson, K., Crutzen, P. and Foley, J. (2009) 'Planetary boundaries: Exploring the safe operating space for humanity', *Ecology and Society,* 14(2): 1–32.

Rodger, N.A.M. (1988) *The Wooden World: An Anatomy of the Georgian Navy.* London: Fontana.

Rodger, N.A.M. (2004) *The Command of the Ocean: A Naval History of Britain, 1649–1815.* London: Allen Lane.

Rogers, B. (1998) 'From Mocha Dick to Moby Dick: Fishing for Clues to Moby's Name and Color', *Names: A Journal of Onomastics,* 46(4): 263–276.

Rollin, B.E. (1981) 'The metaphysics of anthropocentrism: A review of Paul Ehrenfeld's "The Arrogance of Humanism" and Mary Midgley's "Beast and Man", *International Journal for the Study of Animal Problems*, 2(2): 75–80.

Rothenberg, D. (2008) *Thousand-Mile Song: Whale Music in a Sea of Sound*. New York: Basic Books.

Russell, E. (2004) 'The Garden in the Machine: Toward an Evolutionary History of Technology', in Schrepfer, S. and Scranton, P. (Eds) *Industrializing Organisms: Introducing Evolutionary History*. New York: Routledge, 1–18.

Samuels, A. and Tyack, P.L. (2000) 'Flukeprints: A History of Studying Cetacean Societies', in Mann, J., Connor, R.C., Tyack, P.L. and Whitehead, H. (Eds) *Cetacean societies: field studies of dolphins and whales*. Chicago: University of Chicago Press, 9–44.

Savelle, J.M. (1997) 'Architectural Utility in the Formation of Zooarchaeological Whale Bone Assemblages', *Journal of Archaeological Science*, 24(10): 869–885.

Schmidt, A. (1971) *The Concept of Nature in Marx*. London: New Left Books.

Schrepfer, S. and Scranton, P. (Eds) (2004) *Industrializing Organisms: Introducing Evolutionary History*. New York: Routledge.

Schwägerl, C. (2014) *The Anthropocene: The Human Era and How It Shapes our Planet*. Santa Fe: Synergetic Press.

Schwartz, S.L. (1983/1984) 'Sea and Land Symbolism in Mrs Gaskell's *Sylvia's Lovers*', *Estudos Anglo-Americanos*, 7(8): 1–15.

Scoresby, W. (1823) *Journal of a Voyage to the Northern Whale-Fishery*. Edinburgh: Archibald Constable.

Scoresby, W. (1969a) *An Account of the Arctic Regions with a History and Description of the Northern Whale-Fishery, Volume One: The Arctic*. Newton Abbott: David & Charles.

Scoresby, W. (1969b) *An Account of the Arctic Regions with a History and Description of the Northern Whale-Fishery, Volume Two: The Whale-Fishery*. Newton Abbott: David & Charles.

Scoresby, W. (1984) *The 1823 Log Book of the Ship Baffin from Liverpool*. Whitby: Caedmon.

Scoresby, W. (1985) *My Father: Being Records of the Adventurous Life of the Late William Scoresby. Esq. of Whitby*. Whitby: Caedmon.

Semple, E.C. (1911) *Influence of the Geographic Environment*. New York: Henry Holt.

Shackleton, E. (1998) *South: A Memoir of the* Endurance *Voyage*. London: Robinson.

Shukin, N. (2009) *Animal Capital: Rendering Life in Biopolitical Times*. Minneapolis: University of Minnesota Press.

Simmel, G. (1969) 'The Metropolis and Mental Life'. Trans. H.H. Gerth and C. Wright Mills, in Sennett, R. (Ed.) *Classic Essays on the Culture of Cities*. New Jersey: Prentice Hall, 47–60.

Smirnov, A. (2013) *Sound in Z: Experiments in Sound and Electronic Music in Early 20th Century Russia*. London: Koenig Books.

Smol, J.P. (2008) *Pollution of Lakes and Rivers: A Paleoenvironmental Perspective, 2nd Edition*. Oxford: Blackwell.

Sorenson, J. (Ed.) (2014) *Critical Animal Studies: Thinking the Unthinkable*. Toronto: Canadian Scholars' Press.

Souvatzi, S.G. (2008) *A Social Archaeology of Households in Neolithic Greece: An Anthropological Approach*. Cambridge: Cambridge University Press.

Spivak, G. (1988) 'Can the Subaltern Speak?', in Nelson, C. and Grossberg, L. (Eds) *Marxism and the Interpretation of Culture*. Chicago: University of Illinois Press, 271–313.

Stamp, C. (2007) *Scoresby*. Whitby: Caedmon.

Stamp, T. and Stamp, C. (1975) *William Scoresby: Arctic Scientist*. Whitby: Caedmon.

Stamp, T. and Stamp, C. (1983) *Greenland Voyager*. Whitby: Caedmon.

Steiner, G. (1972) *Extra-Territorial: Papers on Literature and the Language Revolution*. London: Faber & Faber.

Stiegler, B. (1998) *Technics and Time, 1: The Fault of Epimetheus*. Trans. R. Beardsworth, and G. Collins. Stanford: Stanford University Press.

Stiegler, B. (2008) *Technics and Time, 2: Disorientation*. Trans. S. Barker. Stanford: Stanford University Press.

Stiegler, B. (2010) *Technics and Time, 3: Cinematic Time and the Question of Malaise*. Trans. S. Barker. Stanford: Stanford University Press.

Stiegler, B. (2010) 'Pharmacology of Desire: Drive-based Capitalism and Libidinal Dis-economy – Keynote lecture delivered to the Psychoanalysis, Money and the Economy Conference, Freud Museum and Birkbeck College, 3 July 2010'. Trans. D. Ross, *New Formations*, 72, 150–161.

Stiegler, B. (2014) *The Re-Enchantment of the World: The Value of Spirit Against Industrial Populism*. Trans. T. Arthur. London: Bloomsbury.

Stoner, A. and Melathopolous, A. (2015) *Freedom in the Anthropocene: Twentieth-Century Helplessness in the Face of Climate Change*. New York: Palgrave Macmillan.

Straus, L.G. (2009) 'The Late Upper Paleolithic-Mesolithic-Neolithic Transitions in Cantabrian Spain', *Journal of Anthropological Research*, 65(2): 287–298.

Swyngedouw, E. (1996) 'The City as a Hybrid', *Capitalism, Nature, Socialism*, 7(2): 65–80.

Szabo, V.E. (2005) '"Bad to the bone?" The Unnatural History of Monstrous Medieval Whales', *The Heroic Age: A Journal of Early Medieval Northwestern Europe*, 8, www.heroicage.org/issues/8/szabo.html [Accessed 7 November 2016].

Szabo, V.E. (2008) *Monstrous Fishes and the Mead-Dark Sea: Whaling in the Medieval North Atlantic*. Leiden: Brill.

Taylor, R. (1911) *The Political Prophecy in England*. New York: Columbia University Press.

Taylor, N. and Twine, R. (Eds) (2014) *The Rise of Critical Animal Studies: From the Margins to the Centre*. London: Routledge.

Thewissen, J.G.M and Fish, F.E. (1997) 'Locomotor evolution in the earliest cetaceans: functional model, modern analogues, and paleontological evidence', *Paleobiology*, 23(4): 482–490.

Thompson, E.P. (1968) *The Making of the English Working Class*. Harmondsworth: Penguin.

Thompson, E.P. (1976) *Whigs and Hunters: The Origin of the Black Act*. New York: Pantheon.

Thompson, P., Wailey, T. and Lummis, T. (1983) *Living the Fishing*. London: Routledge and Kegan Paul.

Tresch, J. (2012) *The Romantic Machine: Utopian Science and Technology After Napoleon*. Chicago: Chicago University Press.

Trotsky, L. (1991) *Literature and Revolution*. Trans. R. Strunsky. London: RedWords.

Truitt, E.R. (2015) *Medieval Robots; Mechanism, Magic, Nature, and Art*. Philadelphia: University of Pennsylvania Press.

Tyack, P.L. (2000) 'Functional Aspects of Cetacean Communication', in Mann, J., Connor, R.C., Tyack, P.L. and Whitehead, H. (Eds) *Cetacean societies: field studies of dolphins and whales.* Chicago: University of Chicago Press, 270–307.

Vaccari, A. and Barnet, B. (2009) 'Prolegomena to a Future Robot History: Stiegler, Epiphylogenesis and Technical Evolution', *Transformations*, 17: 1–16.

Van der Post, L. (1970) *The Hunter and the Whale.* Harmondsworth: Penguin.

Virilio, P. (2006) *Speed and Politics.* Trans. M. Polizzotti. Los Angeles: Semiotext(e).

Wadiwel, D.J. (2015) *The War against Animals.* Leiden: Brill.

Weber, M. (1930) *The Protestant Ethic and the Spirit of Capitalism.* Trans. T. Parsons. New York: The Citadel Press.

Weber, M. (1968) *Economy and Society.* Ed. and Trans. by G. Roth and C. Wittich. New York: Bedminster Press.

Weber, M. (2001) *The Protestant Ethic and the Spirit of Capitalism.* Trans. T. Parsons. London: Routledge.

Weber, M. (2003) *General Economic History.* Trans. F. Knight. Mineola, New York: Dover Publications.

Weissman, S. (2001) *Victor Serge: The Course is Set on Hope.* London: Verso.

Weitzenfeld, A. and Joy, M. (2014) 'An Overview of Anthropocentrism, Humanism, and Speciesism in Critical Animal Theory', in Nocella, A.J., Sorenson, J., Socha, K. and Matsuoka, A. (Eds) *Defining Critical Animal Studies: An Intersectional Social Justice Approach for Liberation.* New York: Peter Lang, 3–27.

Whitehead, H. (2003) *Sperm Whales: Social Evolution in the Ocean.* Chicago: Chicago University Press.

Whitehead, H., Christal, J. and Tyack, P.L. (2000) 'Studying Cetacean Social Structure in Space and Time: Innovative Techniques', in Mann, J., Connor, R.C., Tyack, P.L. and Whitehead, H. (Eds) *Cetacean societies: field studies of dolphins and whales.* Chicago: University of Chicago Press, 65–89.

Whitehead, H., Reeves, R.R. and Tyack, P.L. (2000) 'Science and the Conservation, Protection, and Management of Wild Cetaceans', in Mann, J., Connor, R.C., Tyack, P.L. and Whitehead, H. (Eds) *Cetacean societies: field studies of dolphins and whales.* Chicago: University of Chicago Press, 308–332.

Whitelaw, M. (2004) *Metacreation: Art and Artificial Life.* Cambridge: The MIT Press.

Whitridge, P. (2002) 'Social and ritual determinants of whale bone transport at a classic Thule winter site in the Canadian Arctic', *International Journal of Osteoarchaeology*, 12(1): 65–75.

Williams, A. (2004) 'Disciplining animals: sentience, production, and critique', *International Journal of Sociology and Social Policy*, 24(9): 45–57.

Williams, H. (1988) *Whale Nation.* London: Jonathan Cape.

Williams, M., Zalasiewicz, J., Haff, P.K., Schwägerl, C., Barnosky, A.D. and Ellis, E. C. (2015) 'The Anthropocene Biosphere', *The Anthropocene Review*, 2(3): 196–219.

Wills, D. (1995) *Prosthetics.* Stanford: Stanford University Press.

Wills, D. (2006) 'Techneology or the Discourse of Speed', in Smith, M. and Morra, J. (Eds) *The Prosthetic Impulse: From a Posthuman Present to a Biocultural Future.* Cambridge: MIT Press, 237–263.

Wills, D. (2008) *Dorsality: Thinking Back through Technology and Politics.* Minneapolis: University of Minnesota Press.

Wills, D. (2016) *Inanimation: Theories of Inorganic Life.* Minneapolis: University of Minnesota Press.

Wilson, E. (1966) *Diary of the "Discovery" Expedition to the Antarctic Regions 1901–1904*. London: Blandford Press.

Wilson, P.L. (2003) *Pirate Utopias: Moorish Corsairs and European Renegadoes.* New York: Autonomedia.

Wolmar, C. (2004) *The Subterranean Railway.* London: Atlantic.

Wolmar, C. (2006) *Fire and Steam*. London: Atlantic.

Wolmar, C. (2008) *Blood, Iron and Gold.* London: Atlantic.

Wolmar, C. (2010) *Engines of War.* London: Atlantic.

Woolgar, S. (1985) 'Why not a Sociology of Machines? The Case of Sociology and Artificial Intelligence', *Sociology*, 19(4): 557–572.

Wyschogrod, E. (1990) *Spirit in Ashes: Hegel, Heidegger and Man-Made Mass Death*. New Haven: Yale University Press.

Young, J.Z. (1962) *The Life of Vertebrates, 2nd Edition*. Oxford: Oxford University Press.

Index

www.ingramcontent.com/pod-product-compliance
Ingram Content Group UK Ltd.
Pitfield, Milton Keynes, MK11 3LW, UK
UKHW011015180926
475701UK00004B/753
9780367208202